Springer Tracts in Modern Physics
Volume 182

Springer-Verlag Berlin Heidelberg GmbH

Physics and Astronomy ONLINE LIBRARY

http://www.springer.de/phys/

Springer Tracts in Modern Physics

Springer Tracts in Modern Physics provides comprehensive and critical reviews of topics of current interest in physics. The following fields are emphasized: elementary particle physics, solid-state physics, complex systems, and fundamental astrophysics.

Suitable reviews of other fields can also be accepted. The editors encourage prospective authors to correspond with them in advance of submitting an article. For reviews of topics belonging to the above mentioned fields, they should address the responsible editor, otherwise the managing editor.

See also http://www.springer.de/phys/books/stmp.html

Managing Editor

Gerhard Höhler

Institut für Theoretische Teilchenphysik
Universität Karlsruhe
Postfach 69 80
76128 Karlsruhe, Germany
Phone: +49 (7 21) 6 08 33 75
Fax: +49 (7 21) 37 07 26
Email: gerhard.hoehler@physik.uni-karlsruhe.de
http://www-ttp.physik.uni-karlsruhe.de/

Elementary Particle Physics, Editors

Johann H. Kühn

Institut für Theoretische Teilchenphysik
Universität Karlsruhe
Postfach 69 80
76128 Karlsruhe, Germany
Phone: +49 (7 21) 6 08 33 72
Fax: +49 (7 21) 37 07 26
Email: johann.kuehn@physik.uni-karlsruhe.de
http://www-ttp.physik.uni-karlsruhe.de/~jk

Thomas Müller

Institut für Experimentelle Kernphysik
Fakultät für Physik
Universität Karlsruhe
Postfach 69 80
76128 Karlsruhe, Germany
Phone: +49 (7 21) 6 08 35 24
Fax: +49 (7 21) 6 07 26 21
Email: thomas.muller@physik.uni-karlsruhe.de
http://www-ekp.physik.uni-karlsruhe.de

Fundamental Astrophysics, Editor

Joachim Trümper

Max-Planck-Institut für Extraterrestrische Physik
Postfach 16 03
85740 Garching, Germany
Phone: +49 (89) 32 99 35 59
Fax: +49 (89) 32 99 35 69
Email: jtrumper@mpe-garching.mpg.de
http://www.mpe-garching.mpg.de/index.html

Solid-State Physics, Editors

Hidetoshi Fukuyama
Editor for The Pacific Rim

University of Tokyo
Institute for Solid State Physics
5-1-5 Kashiwanoha
Kashiwa-shi, Chiba 277-8581, Japan
Email: fukuyama@issp.u-tokyo.ac.jp
http://www.issp.u-tokyo.ac.jp/index_e.html

Andrei Ruckenstein
Editor for The Americas

Department of Physics and Astronomy
Rutgers, The State University of New Jersey
136 Frelinghuysen Road
Piscataway, NJ 08854-8019, USA
Phone: +1 (732) 445 43 29
Fax: +1 (732) 445-43 43
Email: andreir@physics.rutgers.edu
http://www.physics.rutgers.edu/people/pips/
Ruckenstein.html

Peter Wölfle

Institut für Theorie der Kondensierten Materie
Universität Karlsruhe
Postfach 69 80
76128 Karlsruhe, Germany
Phone: +49 (7 21) 6 08 35 90
Fax: +49 (7 21) 69 81 50
Email: woelfle@tkm.physik.uni-karlsruhe.de
http://www-tkm.physik.uni-karlsruhe.de

Complex Systems, Editor

Frank Steiner

Abteilung Theoretische Physik
Universität Ulm
Albert-Einstein-Allee 11
89069 Ulm, Germany
Phone: +49 (7 31) 5 02 29 10
Fax: +49 (7 31) 5 02 29 24
Email: steiner@physik.uni-ulm.de
http://www.physik.uni-ulm.de/theo/theophys.html

Andreas Rosenauer

Transmission Electron Microscopy of Semiconductor Nanostructures

An Analysis of Composition and Strain State

With 136 Figures, Including 24 in Color

 Springer

Dr. Andreas Rosenauer

University Karlsruhe
Lab. Electron Microscopy
Engesser Str. 7
76128 Karlsruhe
Germany
E-mail: andreas.rosenauer@lem.uni-karlsruhe.de

Cataloging-in-Publication Data applied for
A catalog record for this book is available from the Library of Congress.

Bibliographic information published by Die Deutsche Bibliothek
Die Deutsche Bibliothek lists this publication in the Deutsche Nationalbibliografie; detailed bibliographic data is available in the Internet at http://dnb.ddb.de.

Physics and Astronomy Classification Scheme (PACS):
68.65.Fg, 68.65.Hb, 68.37.Lp, 61.14.Nm, 68.35.Dv, 68.55.Nq

ISSN print edition: 0081-3869
ISSN electronic edition: 1615-0430

ISBN 978-3-662-14618-7   ISBN 978-3-540-36407-8 (eBook)
DOI 10.1007/978-3-540-36407-8

http://www.springer.de

© Springer-Verlag Berlin Heidelberg 2003
Originally published by Springer-Verlag Berlin Heidelberg New York in 2003
Softcover reprint of the hardcover 1st edition 2003

Typesetting: Camera-ready copy from the author using a Springer LaTeX macro package
Cover concept: eStudio Calamar Steinen
Cover production: *design & production* GmbH, Heidelberg

Printed on acid-free paper   56/3141/YL   5 4 3 2 1 0

Dedicated to the memory of my father,
Adolf Rosenauer

Preface

The high perfection of crystal growth techniques, for example molecular-beam epitaxy (MBE) and the different variants of chemical vapor deposition (CVD), allow the growth of low-dimensional semiconductor heterostructures, which are one of the main research topics in solid-state physics at present. Thin layers or dots of sphalerite-type compound semiconductors such as $Al_xGa_{1-x}As$, $In_xGa_{1-x}As$ and $Cd_xZn_{1-x}Se$ are used in the fabrication of optoelectronic devices such as light-emitting diodes and laser diodes. The optical and electronic properties of such devices are strongly influenced by the local variation of the composition x because it affects the positions and spacings of the discrete energy levels of charge carriers in quantum wells and dots.

However, there is a lack of basic understanding of the growth processes, in particular for the three-dimensional growth modes, where isolated islands are nucleated on a continuous wetting layer covering the substrate (Stranski-Krastanov growth mode) or directly on the substrate (Volmer-Weber growth mode). The three-dimensional growth modes are used to obtain self organized nanostructures, referred to as "quantum dots". The optical and electronic properties are being studied intensively by many groups; correlation of these properties with the structural and compositional properties is required. In this context, effects such as segregation, interdiffusion and strain-driven adatom migration must be investigated with an atomic-scale spatial resolution.

However, the evaluation of specimen properties such as the local composition is a difficult problem. First, the interaction of a thin object with an electron beam mainly affects the local phase of the electrons, which unfortunately is lost in the imaging process, where only the intensity is recorded. Second, lens aberrations and the possible defocusing of the objective lens lead to a blurring of the information. Quantitative determination of the morphology and chemical composition of semiconductor nanostructures requires the application of optimized imaging and evaluation methods. Although many books are available that deal with the fundamentals and theoretical aspects of transmission electron microscopy, there is a lack of books that describe practical methods for obtaining properties of semiconductor nanostructures from transmission electron microscope images. One goal of this book is to give basic information about digital evaluation techniques for high resolution-transmission electron microscope images. The book also contains the author's own developments and results and describes readily applicable tools for composition determination in ternary semiconductor nanostructures with a near-atomic-scale spatial resolution, focusing on new methods, including strain state analysis and the technique of composition evaluation by lattice fringe analysis.

The work described here was performed at the Laboratory for Electron Microscopy at the University of Karlsruhe, and at the Electron Microscopy for Material Research (EMAT) group at the University of Antwerp. It was partly funded by the Volkswagen Foundation, the Deutsche Forschungsgesellschaft, the Alexander-von-Humboldt Foundation and the Interuniversity Poles of Attraction Program. I want to acknowledge the support and collaboration of many friends and colleagues. These include Prof. Dr. D. Gerthsen, who encouraged me to write this book, for her great friendly support and many fruitful discussions. I thank Prof. Dr. Dirk Van Dyck for enabling my stay in Antwerp and for the explanation of channeling theory, Prof. Dr. D. Schreyvers for his help in installing the electron biprism, and Prof. Dr. H. Lichte, whose group manufactured the biprism. I am grateful to Dr. G. Lang for his explanations in Tübingen and to Dr. M. Lehmann for many valuable hints. I thank S. Kalhöfer and A. Kamilli for the preparation of numerous specimens, Dr. A. Förster of the Forschungszentrum Jülich for the MBE epitaxy of many samples, and for the same reason Dr. G. Böhm of the Walter-Schottky Institute in Garching, as well as Dr. M. Arzberger for his photoluminescence investigations and for interesting and valuable discussions. I am grateful to Dr. K. Tillmann and Dr. A. Thust for their illuminating comments. I thank Prof. Dr. D. Gerthsen, Dipl. Phys. P. Kruse and Dipl. Phys. M. Schowalter for the careful reading of the manuscript, and my collegues Dipl. Phys. Thilo Remmele, Dipl. Phys. A. Wurl, Dipl. Phys. U. Fischer, Dipl. Phys. W. Oberst, Dipl. Phys. N. Peranio, Dipl. Phys. D. Litvinov, Dr. B. Neubauer, Dipl. Phys. P. Kruse, Dipl. Phys. M. Schowalter and Dr. Ana-Mercedes Castellanos-Almonacid for their contributions. I thank Prof. Dr. J. Zweck, who taught me to question everything, for his multifaceted support, R. Sauter for her steady helpfulness in questions concerning "bureaucratic" problems, and Dipl. Phys. W. Send for solving many technical problems. I thank my wife Ulrike and my daughter Lea for their patience during my weekend writing of the manuscript. I thank the EMAT group for their hospitality and the felicitous Christmas party. Finally, I thank all members of the LEM group for their support and their friendly atmosphere.

Karlsruhe, August 2002   *Andreas Rosenauer*

Contents

Part II Digital Image Analysis

Part III Applications

1 Introduction

The techniques described in this book are aimed at investigation of the properties of crystal with a spatial resolution that provides the view into the unit cell of the crystal. For this purpose, high-energy electrons are preferable, among all the different kinds of radiation that could be used. First, electrons possess a charge, and a beam of electrons can be focused in an inhomogeneous magnetic field, which allows the construction of an electromagnetic focusing lens. Among charged particles, electrons (and also positrons, but these are not useful here) possess the smallest mass, which minimizes the structural damage that they cause in the specimen. In the transmission electron microscope (TEM), electrons are accelerated to a few hundreds of keV. The de Broglie wavelength of the electrons is of the order of only a few picometers, and the point resolution of modern TEMs lies in the 0.1 nm range. In addition to the good spatial resolution, the strong interaction of the electrons with matter allows the interaction volume to be extremely small. One single column of only a few atoms is sufficient to determine the positions and, in principle, also the types of the atoms from the scattered electron wave.

The interaction of the electron beam with the specimen provides many channels of information that can be used for compositional analysis. First, the inelastic scattering of electrons can be used for energy-dispersive X-ray analysis (EDX), electron energy loss spectroscopy (EELS) and energy-filtered TEM (EFTEM) [1]. The spatial resolution, of the order of a few nanometers, is generally not sufficient to measure the composition of nanostructures with good accuracy. As an alternative, high-resolution TEM (HRTEM) can be utilized in combination with appropriate image evaluation techniques, where a resolution of the order of 0.2 nm can be achieved.

"Chemical lattice imaging" was introduced by Ourmazd and coworkers [2]; This allowed the chemical analysis of $Al_xGa_{1-x}As$ heterostructures on an atomic scale. Another example of an evaluation method related to HRTEM is the QUANTITEM (quantitative analysis of information from transmission electron micrographs) procedure which yields the projected potential of the samples for crystals such as Si and Ge when only two Bloch waves are strongly excited [3, 4, 5]. An alternative method, by Stenkamp and Jäger [6], uses systematic variations of the image contrast pattern to obtain the local composition of SiGe alloys for certain ranges of objective lens defocus Δf and specimen thickness t by measuring local Fourier coefficients.

This book focuses on two methods to quantify the information contained in HRTEM images that are particularly useful for the investigation of semi-conductor heteroepitaxial layers, where the strain state and the composition on an atomic scale are of interest: strain state analysis, and composition evaluation by lattice fringe analysis (CELFA).

1.1 Strain State Analysis [7, 8, 9, 10]

The basis of one possible approach to the problem of determining strain and composition on an atomic scale is the measurement of local lattice parameters, i.e. the measurement of distances between adjacent atomic columns. This approach is based principally on an important result of channeling theory: presupposing an electron beam parallel to a zone axis of the specimen, Van Dyck et al. showed [11] that the positions of the atomic columns are given by the local intensity maxima of the electron wave function at the object exit surface. If the objective lens introduces only radially symmetric aberrations, this relation holds even in the image plane for perfect-crystal specimens. Nevertheless, the objective lens aberrations give rise to delocalization. The information content of each point of the wave function at the object exit surface is spread over an area in the image plane whose size depends on the lens aberrations and defocus. Consequently, sharp interfaces appear blurred in the image. These effects can be minimized by using the proper defocus. Although errors have to be expected close to chemical transitions in the specimen and in regions where the specimen thickness changes rapidly, the positions of maximum image intensity are well suited for the measurement of local lattice parameter variations and of displacements of atomic columns with good accuracy. For this purpose, only a sufficiently constant relationship needs to be assumed between the positions of the atomic columns and the positions of the intensity maxima, whereas the relationship itself does not need to be known at all.

The impact of objective lens aberrations can be avoided by the reconstruction of the aberration-free exit wave function, which can be achieved by two different approaches. In one of these approaches, a Möllenstedt biprism is inserted close to the first intermediate image plane to perform off-axis electron holography [12, 13]. In the other approach, a series of HRTEM images taken at systematically varied defocus values is used in the focal-variation reconstruction method published by Coene et al. [14].

The local composition can be extracted from a single HRTEM image or, more accurately, from the amplitude of the reconstructed electron exit wave function, if the relationship between the composition and the lattice parameter is known. For many compound semiconductors, e.g. $In_xGa_{1-x}As$ and $Cd_xZn_{1-x}Se$, Vegard's law,

$$a_{A_xB_{1-x}C} = a_{BC} + x(a_{AC} - a_{BC}) \, , \qquad (1.1)$$

can be applied; here the lattice parameter and the composition are linearly correlated. If a mismatch exists between the lattice parameters of the substrate and the epilayer, the distortion of the unit cells in the epilayer must be taken into account. The tetragonal distortion can be easily calculated for coherently strained two-dimensional layers below the critical thickness [15] for plastic relaxation by misfit dislocations.

Measurement of local lattice parameters was applied by Bierwolf et al. [16] and Jouneau et al. [17] to investigate the strain distribution of thin epitaxial layers. Robertson et al. [18] used Fourier-filtered HRTEM images to measure the spacings and cumulative deviations of lattice fringes. Hÿtch et al. [19, 20, 21] used a Fourier filtering technique to derive the geometric local phase of reflections in the Fourier-transformed image (diffractogram).

The situation becomes more difficult for three-dimensional growth modes. Deviations from a tetragonal distortion occur close to the surface owing to the elastic relaxation of the strained lattice. The finite-element method (FEM) was first applied to compute the strain distribution in nanoscale islands in the SiGe/Si(001) by Christiansen et al. [22] . A complete relaxation of the misfit strain close to the island surface was obtained; this is regarded as the major driving force for the island growth. Therefore, an accurate knowledge of the strain distribution is a necessary prerequisite for composition evaluation from local lattice parameters in the case of epitaxial islands.

Another question to be addressed is the elastic relaxation of strained structures due to the small thickness of an HRTEM specimen (typically 20 nm), which can significantly modify the tetragonal distortion, depending on the local specimen thickness and the dimensions of the strained structure. The specimen thickness must be accurately measured in the region of interest of the HRTEM image that is being analyzed. A further important step in the quantification is the calculation of the elastic relaxation as a function of the thickness of the TEM specimen and the layer morphology. Analytical solutions to this problem for simple layer morphologies have been presented by Treacy et al. [23]. For more complicated morphologies, FEM simulations can be applied [24].

1.2 CELFA [7, 25, 26, 27, 28]

Owing to the modification of local lattice parameters by the local specimen thickness and by elastic strain relaxation in islands, a different and less elaborate approach to composition determination is desirable. In this book, an alternative method is described that exploits chemically sensitive reflections such as the {002} reflections available in sphalerite-type crystals. This type of reflection provides an amplitude that depends strongly on the crystal composition but often is rather small, as in $In_xGa_{1-x}As$ and $Cd_xZn_{1-x}Se$, for example.

Single-beam dark-field imaging with a chemically sensitive reflection is conventionally used to display variations of the chemical composition qualitatively. Since the local image intensity is proportional to the square of the amplitude of the chemically sensitive reflection, quantitative data can be extracted, in principle. However, several problems exist. The signal-to-noise ratio is typically small, owing to the small amplitude of the chemically sensitive reflection and to inelastic scattering. Another significant disadvantage may be explained as follows. The structure amplitude of a chemically sensitive reflection depends linearly on the elemental concentration x in a semiconductor $A_x B_{1-x} C$. Depending on the atomic scattering factors of the atoms involved, the sign of the structure amplitude may change at a certain concentration x_0. As a consequence, the amplitudes of the chemically sensitive beam are similar at values of x equal to $x_0 \pm \delta x$ whereas the phases differ by π. It is a disadvantage of the single-beam dark-field imaging technique that the phase of the beam is lost, but is essential to resolve the ambiguity described above.

The CELFA technique leads to a significantly improved signal-to-noise ratio and recovers the phase information contained in the chemically sensitive beam. The technique exploits a two- or three-beam interference of the chemically sensitive beam with the undiffracted beam. A three-beam condition using an additional, third reflection can be used to obtain information about the local specimen thickness. An off-axis imaging condition is used to enlarge the extinction distance, which minimizes the influence of specimen thickness variations. A particularly elegant way to obtain the amplitude and phase of the chemically sensitive beam, as well as the local specimen thickness, is off-axis electron holography, where the chemically sensitive reflection interferes with a reference beam that is spatially homogeneous. The specimen thickness is evaluated from the phase of the central beam of the centered sideband of the electron hologram.

The methods described above have been applied to a variety of material systems, including $In_x Ga_{1-x} As$ [7, 24, 29, 30, 31, 32, 33, 34], $Cd_x Zn_{1-x} Se$ [7, 35, 36, 37, 38, 39, 40, 41, 42, 43, 44, 45, 46, 47, 48, 49, 50, 51, 52], $Al_x Ga_{1-x} As$ [25, 53, 54], $Al_x Ga_{1-x} N$ [55, 56] and $In_x Ga_{1-x} N$ [57, 58, 59, 60]. A discussion of all these applications clearly goes beyond the scope of this book. Therefore, the presentation of applications will be restricted to investigations of $In_x Ga_{1-x} As$ Stranski-Krastanov (SK) layers and composition determination of $Al_x Ga_{1-x} As$ / GaAs superlattices by off-axis electron holography.

1.3 Organization of the Book

The present book is organized in the following way. The first part provides the theoretical fundamentals of transmission electron microscopy needed in the second part, which focuses on a description of strain state analysis and on the composition evaluation by lattice fringe analysis techniques. In the third part, we describe the application of these techniques to the investigation of low-dimensional semiconductor heterostructures such as $In_xGa_{1-x}As$ SK layers.

In Part I of this book, we describe the electron wave on its way from the specimen surface to the image. Chapter 2 starts with the interaction of the electron wave with a crystalline specimen. First, we treat the scattering by a single atom in Sect. 2.1. Section 2.2 describes the effect of electron diffraction in the kinematical approximation, which is valid for crystals with a thickness of only a few nanometers in the direction of the electron beam. More realistic specimen thicknesses are treated in Sect. 2.3 within the framework of the Bloch wave approach. Finally, channeling theory is used in Sect. 2.4 to show that the positions of maximum intensity of the wave function at the object exit surface correspond to the positions of atomic columns, presupposing an exact zone axis orientation of the specimen.

Chapter 3 is concerned with the intensity pattern observed in the image plane. It starts with the fictitious assumption of an ideal microscope and then allows for spherical aberration and defocus in Sect. 3.1. Effects of incoherence such as fluctuations of the high tension or objective lens current are treated in Sect. 3.2. At the end of Chap. 3, two approaches are described in Sect. 3.3 that allow the reconstruction of the wave function at the object exit surface: the focal-variation technique and off-axis electron holography.

Part II deals with the methods that have been developed for digital image analysis. The first procedure, discussed in Chap. 4, is strain state analysis. Section 4.1 outlines the measurement of the displacements and spacings of lattice positions. A knowledge of the local specimen thickness is an important prerequisite because the tetragonal distortion in a thin TEM specimen is reduced in comparison with a bulk sample. Section 4.2 outlines a procedure to measure the thickness, based upon QUANTITEM. After that, Sect. 4.3 focuses on the determination of the elastic relaxation of a thin specimen by FEM simulations.

A rather detailed description of the CELFA technique is presented in Chap. 5. The basic ideas behind CELFA are introduced in Sect. 5.1. Subsequently, Sect. 5.2 gives a theoretical treatment of the technique. Practical considerations follow in Sect. 5.3, which is concerned with the actual procedures for the measurement of amplitudes and phases of reflections, as well as with the errors in the evaluated composition due to objective lens aberrations and specimen thickness uncertainties. The effect of strain is addressed in Sect. 5.4 and, finally, the impact of a nonrandom distribution of atom types that share the same crystal sublattice is discussed in Sect. 5.5.

Applications of the evaluation methods are given in Part III. Chapter 6 introduces the Stranski-Krastanov growth mode (Sect. 6.1) and gives a survey of the present level of understanding of segregation effects in III-V ternary alloys (Sect. 6.2). Chapter 7 outlines the investigations of $In_{0.6}Ga_{0.4}As$ SK layers. It shows how the application of strain state analysis and the CELFA technique allows novel insights into the morphology of free-standing and capped SK layers. The investigations reveal that effects such as strain-induced migration of Ga and In, segregation, and incorporation of migrating In into the growing cap layer lead to a considerable morphological transformation of the SK layer during overgrowth of GaAs.

The effect of segregation leads to interesting and surprising morphologies of the wetting layer and islands as has been shown by the investigation of nominally binary InAs quantum dots presented in Chap. 8. Here we have found that the putative binary islands contained more than 50% Ga. An investigation of the wetting layers revealed, in a very clear and unambiguous manner, the existence of segregation. This observation is in contradiction to the wide-spread assumption that segregation is based upon an exchange reaction of In and Ga at the growth surface.

Chapter 9 makes the point that electron holography could be very useful for measuring the composition of materials in cases where the specimen thickness is of crucial importance in chemical analysis. Here we show that the measurement of the local phase of the (000) beam of the centered sideband, combined with the amplitude of the chemically sensitive (002) beam, allows one to deduce both the specimen thickness and the composition, in an iterative and self-consistent way. Although accurate values of the mean inner potential are generally not available at present, the suggested method leads to good accuracy in the measured composition. Here we describe the application of the method to an AlAs/GaAs superlattice, where again the effect of segregation is demonstrated and its efficiency is measured.

References

1. W. Grogger, F. Hofer, P. Warbichler, G. Kothleitner: Microsc. Microanal. **6**, 161 (2000)
2. A. Ourmazd, D.W. Taylor, J. Cunningham, C.W. Tu: Phys. Rev. Lett. **62**, 933 (1989)
3. C. Kisielowski, P. Schwander, F.H. Baumann, M. Seibt, Y.O. Kim, A. Ourmazd: Ultramicroscopy **58**, 131 (1995)
4. A. Ourmazd, P. Schwander, C. Kisielowski, M. Seibt, F.H. Baumann, Y.O. Kim: Inst. Phys. Conf. Ser. **134**: Section 1, 1 (1993)
5. J.-L. Maurice, P. Schwander, F.H. Baumann, A. Ourmazd: Ultramicroscopy **68**, 149 (1997)
6. D. Stenkamp, W. Jäger: Ultramicroscopy **50**, 321 (1993)
7. A. Rosenauer, D. Gerthsen: Adv. Imaging Electron Phys. **107**, 121 (1999)
8. A. Rosenauer, T. Remmele, U. Fischer, A. Förster, D. Gerthsen: "Strain determination in mismatched semiconductor heterostructures by the digital analysis of lattice images". In: *Microscopy of Semiconducting Materials 1997*, Oxford, UK, April 7–10, 1997, Proceedings of the Royal Microscopical Society Conference, ed. by A.G. Cullis, J.L. Hutchison (Institute of Physics, Bristol, 1997) pp. 39–42

9. A. Rosenauer, T. Remmele, D. Gerthsen, K. Tillmann, A. Förster: Optik **105**, 99 (1997)
10. A. Rosenauer, S. Kaiser, T. Reisinger, J. Zweck, W. Gebhardt, D. Gerthsen: Optik **102**, 63 (1996)
11. D. Van Dyck, M. Op de Beeck: Ultramicroscopy **64**, 99 (1996)
12. H. Lichte: Ultramicroscopy **47**, 223 (1992)
13. H. Lichte, E. Völkl: Ultramicroscopy **47**, 231 (1992)
14. W.M.J. Coene, A. Thust, M. Op de Beeck, D. Van Dyck: Ultramicroscopy **64**, 109 (1996)
15. R. Hull, J.C. Bean: Critical Reviews in Solid State and Materials Sciences **17**, 507 (1992)
16. R. Bierwolf, M. Hohenstein, F. Phillipp, O. Brandt, G.E. Crook, K. Ploog: Ultramicroscopy **49**, 273 (1993)
17. P.H. Jouneau, A. Tardot, G. Feulliet, H. Marietta, J. Cibert: J. Appl. Phys. **75**, 7310 (1994)
18. M.D. Robertson, J.E. Curie, J.M. Corbett, J.B. Webb: Ultramicroscopy **58**, 175 (1995)
19. M.J. Hytch, L. Potez: Phil. Mag. A **76**, 1119 (1997)
20. M.J. Hytch: Microsc. Microanal. Microstruct. **8**, 41 (1997)
21. M.J. Hytch, E. Snoeck, R. Kilaas: Ultramicroscopy **74**, 131 (1998)
22. S. Christiansen, M. Albrecht, H.P. Strunk, H.J. Maier: Appl. Phys. Lett. **64**, 3617 (1994)
23. M.M.J. Treacy, J.M. Gibson: J. Vac. Sci. Technol. B **4**, 1458 (1986)
24. K. Tillmann, A. Thust, M. Lentzen, P. Swiatek, A. Förster, K. Urban, D. Gerthsen, T. Remmele, A. Rosenauer: Phil. Mag. Lett. **74**, 309 (1996)
25. A. Rosenauer, D. Van Dyck, M. Arzberger, G. Abstreiter: Ultramicroscopy **88**, 51 (2001)
26. A. Rosenauer, D. Van Dyck: "The effect of strain on the chemically sensitive imaging with the (002) reflection in sphalerite type crystals". In: *Proceedings of EUREM 12*, Brno, Czech Republic, July 9–14, 2000, p. I121
27. A. Rosenauer, D. Gerthsen: Ultramicroscopy **76**, 49 (1999)
28. A. Rosenauer, U. Fischer, D. Gerthsen, A. Förster: Ultramicroscopy **72**, 121 (1998)
29. A. Rosenauer, D. Gerthsen, D. Van Dyck, M. Arzberger, G. Böhm, G. Abstreiter: Phys. Rev. B **64**, 245334-1 (2001)
30. A. Rosenauer, W. Oberst, D. Litvinov, D. Gerthsen, A. Foerster, R. Schmidt: Phys. Rev. B **61**, 8276 (2000)
31. A.F. Tsatsulnikov, B.V. Volovik, N.N. Ledentsov, M.V. Maximov, A.Yu. Egorov, A.R. Kovsh, V.M. Ustinov, A.E. Zhukov, P.S. Kopev, Z.I. Alferov, I.A. Kozin, M.V. Belousov, I.P. Soshnikov, P. Werner, D. Litvinov, U. Fischer, A. Rosenauer, D. Gerthsen: J. Electron. Mater. **28**, 537 (1999)
32. A. Rosenauer, W. Oberst, D. Gerthsen, A. Förster: Thin Solid Films **357**, 18 (1999)
33. A. Rosenauer, U. Fischer, D. Gerthsen, A. Förster: Appl. Phys. Lett. **71**, 3868 (1997)
34. U. Woggon, W. Langbein, J.M. Hvam, A. Rosenauer, T. Remmele, D. Gerthsen: Appl. Phys. Lett. **71**, 377 (1997)
35. M. Strassburg, T. Deniozou, A. Hoffmann, S. Rodt, V. Turck, R. Heitz, U.W. Pohl, D. Bimberg, D. Litvinov, A. Rosenauer, D. Gerthsen, S. Schwedhelm, I. Kudryashov, K. Lischka, D. Schikora: J. Cryst. Growth **214–215**, 756 (2000)

36. D. Gerthsen, A. Rosenauer, D. Litvinov, N. Peranio: J. Cryst. Growth **214–215**, 707 (2000)
37. D. Schikora, S. Schwedhelm, I. Kudryashov, K. Lischka, D. Litvinov, A. Rosenauer, D. Gerthsen, A. Strassburg, A. Hoffmann, D. Bimberg: J. Cryst. Growth **214–215**, 698 (2000)
38. M. Strassburg, M. Dworzak, A. Hoffmann, R. Heitz, U.W. Pohl, D. Bimberg, D. Litvinov, A. Rosenauer, D. Gerthsen, I. Kudryashov, K. Lischka, D. Schikora: Phys. Status Solidi A **180**, 281 (2000)
39. N.N. Ledentsov, I.L. Krestnikov, M. Strassburg, R. Engelhardt, S. Rodt, R. Heitz, U.W. Pohl, A. Hoffmann, D. Bimberg, A.V. Sakharov, W.V. Lundin, A.S. Usikov, Z.I. Alferov, D. Litvinov, A. Rosenauer, D. Gerthsen: Thin Solid Films **367**, 40 (2000)
40. D. Litvinov, A. Rosenauer, D. Gerthsen, N.N. Ledentsov: Phys. Rev. B **61**, 16819 (2000)
41. N. Peranio, A. Rosenauer, D. Gerthsen, S.V. Sorokin, I.V. Sedova, S.V. Ivanov: Phys. Rev. B **61**, 16015 (2000)
42. M. Strassburg, T. Deniozou, A. Hoffmann, R. Heltz, U.W. Pohl, D. Bimberg, D. Litvinov, A. Rosenauer, D. Gerthsen, S. Schwedhelm, K. Lischka, D. Schikora: Appl. Phys. Lett. **76**, 685 (2000)
43. D. Schikora, S. Schwedhelm, D.J. As, K. Lischka, D. Litvinov, A. Rosenauer, D. Gerthsen, M. Strassburg, A. Hoffmann, D. Bimberg: Appl. Phys. Lett. **76**, 418 (2000)
44. A. Rosenauer, N. Peranio, D. Gerthsen: "Investigation of CdSe/ZnSe quantum dot structures by composition evaluation by lattice fringe analysis". In: *Microscopy of Semiconducting Materials 1999*, Oxford, UK, March 22–25, 1999, Proceedings of the Institute of Physics Conference, Institute of Physics Conference Series Number 164, ed. by A.G. Cullis, J.L. Hutchison, (Institute of Physics, Bristol, 1997)
45. R. Engelhardt, U.W. Pohl, D. Bimber, D. Litvinov, A. Rosenauer, D. Gerthsen: J. Appl. Phys. **86**, 5578 (1999)
46. I.L. Krestnikov, M. Strassburg, M. Caesar, A. Hoffmann, U.W. Pohl, D. Bimberg, N.N. Ledentsov, P.S. Kopev, Zh.I. Alferov, D. Litvinov, A. Rosenauer, D. Gerthsen: Phys. Rev. B **60**, 8695 (1999)
47. M. Strassburg, R. Heitz, V. Turck, S. Rodt, U.W. Pohl, A. Hoffmann, D. Bimberg, I.L. Krestnikov, V.A. Shchuxin, N.N. Ledentsov, Z.I. Alferov, D. Litvinov, A. Rosenauer, D. Gerthsen: J. Electron. Mater. **28**, 506 (1999)
48. K.G. Chinyama, K.P. O'Donnell, A. Rosenauer, D. Gerthsen: J. Cryst. Growth **203**, 362 (1999)
49. I.L. Krestnikov, S.V. Ivanov, P.S. Kopoev, N.N. Ledentsov, M.V. Maximov, A.V. Sakharov, S.V. Sorokin, A. Rosenauer, D. Gerthsen, C.M. Sotomayor-Torres, D. Bimberg, Zh.I. Alferov: Mater. Sci. Eng. B **51**, 26 (1998)
50. M. Strassburg, A.V. Kutzer, U.W. Pohl, A. Hoffmann, I. Broser, N.N. Ledentsov, D. Bimberg, A. Rosenauer, U. Fischer, D. Gerthsen, I.L. Krestnikov, M.V. Maximov, C.O. Kop: Appl. Phys. Lett. **72**, 942 (1998)
51. T. Reisinger, S. Lankes, M.J. Kastner, A. Rosenauer, F. Franzen, M. Meier, W. Gebhardt: J. Cryst. Growth **159**, 510 (1996)
52. A. Rosenauer, T. Reisinger, E. Steinkirchner, J. Zweck, W. Gebhardt: J. Cryst. Growth **152**, 42 (1995)

53. E. Schomburg, S. Brandl, S. Winnerl, K.F. Renk, N.N. Ledentsov, V.M. Ustinov, A. Zhukov, P.S. Kopev, H.W. Hubers, J. Schubert, H.P. Roser, A. Rosenauer, D. Litvinov, D. Gerthsen, J.M. Chamberlain: Physica E **7**, 814 (2000)

54. E. Schomburg, S. Brandl, K.F. Renk, N.N. Ledentsov, V.M. Ustinov, A.E. Zhukov, A.R. Kovsh, A.Yu. Egorov, R.N. Kyutt, B.V. Volovik, P.S. Kopev, Zh.I. Alferov, A. Rosenauer, D. Litvinov, D. Gerthsen, D.G. Pavelev, Y.I. Koschurinov: Phys. Lett. A **262**, 396 (1999)

55. B. Neubauer, A. Rosenauer, D. Gerthsen, O. Ambacher, M. Stutzmann, M. Albrecht, H.P. Strunk: Mater. Sci. Eng. B **59**, 182 (1999)

56. B. Neubauer, A. Rosenauer, D. Gerthsen, O. Ambacher, M. Stutzmann: Appl. Phys. Lett. **73**, 930 (1998)

57. D. Gerthsen, E. Hahn, B. Neubauer, A. Rosenauer, O. Schön, M. Heuken: Phys. Status Solidi A **177**, 145 (2000)

58. I.P. Soshnikov, V.V. Lundin, A.S. Usikov, I.P. Kalmykova, N.N. Ledentsov, A. Rosenauer, B. Neubauer, D. Gerthsen: Semiconductors **34**, 621 (2000)

59. B. Neubauer, H. Widmann, D. Gerthsen, T. Stephan, H. Kalt, J. Bläsing, P. Veit, A. Krost, O. Schön, M. Heuken: "Structural properties, In distribution and photoluminescence of multiple InGaN/GaN quantum well structures". In: *Proceedings of International Workshop on Nitride Semiconductors*, Nagoya, (Institute of Pure and Applied Physics, Tokyo, 2000) vol. 1002, p. 375

60. D. Gerthsen, B. Neubauer, A. Rosenauer, T. Stephan, H. Kalt, O. Schön, M. Heuken: Appl. Phys. Lett. **79**, 2552 (2001)

Part I

Theoretical Fundamentals
of Transmission Electron Microscopy

2 Electron Diffraction

This chapter describes the elastic Coulomb interaction of an incident electron wave with a crystalline specimen. We start with the scattering of the electron wave function of the incident electrons by a single atom. The integral form of the Schrödinger equation results in a recursive description of the scattered electron wave function Ψ_S as a Born series. The scattered electron wave is calculated in the first Born approximation, and we find that the atomic scattering amplitude for electrons quickly decreases with increasing scattering angle. The periodic arrangement of atoms in the crystalline specimen together with electron scattering leads to electron diffraction. The structure amplitude of the crystal unit cell is used to describe the diffraction of electrons by a thin-foil specimen. This kinematical approximation reveals the origin of the chemically sensitive beams that constitute the basis of composition evaluation by the CELFA method. Dynamical effects of electron diffraction are taken into account using the Bloch wave formalism. With respect to the CELFA technique, we discuss the dependence of the beams on the specimen thickness and show how this dependence can be influenced by varying the excitation condition. Finally, we set out the basis of strain state analysis by discussing the correlation between the atomic positions and the positions of maximum image intensity that appear in the amplitude"image" of the wave function at the exit surface of the object in the framework of channeling theory. Further information about the topics discussed in this chapter may be found in [1, 2, 3].

2.1 Single-Atom Electron Scattering

2.1.1 The Integral Form of the Schrödinger Equation

We start our brief outline of diffraction theory with the scattering of an electron by a single atom that is tightly bound in a crystal. We consider only the case of elastic scattering, where momentum and energy are conserved. Owing to the large difference between the masses of the electron and the crystal, the energy transfer from the electron to the crystal is negligible and the electron wavelength does not change in the scattering process. The problem can be described by the stationary

Schrödinger equation

$$\nabla^2\Psi(\mathbf{r}) + \frac{8\pi^2 me}{h^2}\left(E + \Phi(\mathbf{r})\right)\Psi(\mathbf{r}) = 0 \, , \tag{2.1}$$

where E and m are the relativistically corrected values of the accelerating potential and of the mass of the incident electron, respectively, $\Phi(\mathbf{r})$ is the Coulomb potential of the scattering atom, e is the elementary charge, and h is Planck's constant. Even in an electron microscope operating at only 100 keV, the electrons travel at over half the speed of light. It is thus clear that relativistic corrections must be taken into account. The relativistic corrections involved in (2.1) are

$$E = U\left(1 + \frac{eU}{2m_0 c^2}\right) \, ,$$

$$m = m_0\sqrt{\left(1 + \frac{h^2}{m_0^2 c^2 \lambda^2}\right)} \, ,$$

$$\lambda = \frac{h}{\sqrt{2m_0 eU\left(1 + eU/2m_0 c^2\right)}} \, , \tag{2.2}$$

where U is the accelerating voltage, λ the de Broglie wavelength, m_0 the rest mass of the incident electron and c the speed of light in vacuum. Equation (2.1) can be interpreted as an inhomogeneous differential equation with an inhomogeneity

$$f(\mathbf{r}) := -\frac{8\pi^2 me}{h^2}\Phi(\mathbf{r})\Psi(\mathbf{r}) \, . \tag{2.3}$$

Using the abbreviation

$$k'^2 := \frac{2me}{h^2}E \tag{2.4}$$

and defining the linear differential operator

$$\mathbb{L} := \left[\nabla^2 + 4\pi^2 k'^2\right] \, , \tag{2.5}$$

we can obtain a Green's function

$$G(\mathbf{r} - \mathbf{r}') = -\frac{1}{4\pi}\frac{\exp(-2\pi i k'|\mathbf{r} - \mathbf{r}'|)}{|\mathbf{r} - \mathbf{r}'|} \, , \tag{2.6}$$

which obeys

$$\mathbb{L}\,G(\mathbf{r} - \mathbf{r}') = \delta(\mathbf{r} - \mathbf{r}') \, , \tag{2.7}$$

where δ is Dirac's delta "function"[1] (Sect. B.3). A solution of the inhomogeneous differential equation

---

[1] Equation (2.7) can easily be proven using the relation $\nabla^2(1/|\mathbf{r} - \mathbf{r}'|) = -4\pi\delta(\mathbf{r} - \mathbf{r}')$.

$$\mathbb{L}\Psi(\mathbf{r}) = f(\mathbf{r}) \tag{2.8}$$

is given by

$$\Psi(\mathbf{r}) = \Psi_0(\mathbf{r}) + \int_\Omega G(\mathbf{r} - \mathbf{r}')f(\mathbf{r}')\mathrm{d}^3\mathbf{r}' \, , \tag{2.9}$$

where Ω is the scattering volume. Inserting $G(\mathbf{r} - \mathbf{r}')$ from (2.6), we finally obtain

$$\boxed{\Psi(\mathbf{r}) = \Psi_0(\mathbf{r}) + \frac{2\pi me}{h^2} \int_\Omega \frac{\exp(-2\pi i k'|\mathbf{r} - \mathbf{r}'|)}{|\mathbf{r} - \mathbf{r}'|} \Phi(\mathbf{r}')\Psi(\mathbf{r}')\mathrm{d}^3\mathbf{r}' \, ,} \tag{2.10}$$

where $\Psi_0(\mathbf{r})$ is a solution of the homogeneous differential equation (2.7). An expression for this solution can be obtained from the condition that $\Psi(\mathbf{r}) = \Psi_0(\mathbf{r})$ for $\Phi(\mathbf{r}) = 0$. Thus, $\Psi_0(\mathbf{r})$ is the incident electron wave

$$\Psi_0(\mathbf{r}) = \exp(-2\pi i \mathbf{k}_0 \cdot \mathbf{r}) \, , \tag{2.11}$$

with a wave vector $\mathbf{k}_0$.

2.1.2 The Atomic Scattering Amplitude for Electrons

Using (2.10), $\Psi(\mathbf{r})$ can be expressed as a Born series

$$\Psi(\mathbf{r}) = \sum_{n=0}^{\infty} \Psi_n(\mathbf{r}) \, , \tag{2.12}$$

where $\Psi_n(\mathbf{r})$ is obtained from the integral (2.10) by putting $\Psi(\mathbf{r}) = \Psi_{n-1}(\mathbf{r})$. In the first Born approximation we obtain

$$\Psi_1(\mathbf{r}) = \sigma \int_\Omega \frac{\exp(-2\pi i k'|\mathbf{r} - \mathbf{r}'|)}{|\mathbf{r} - \mathbf{r}'|} \Phi(\mathbf{r}') \exp(-2\pi i \mathbf{k}_0 \mathbf{r}')\mathrm{d}^3\mathbf{r}' \, , \tag{2.13}$$

where σ is the interaction constant, given by

$$\sigma = \frac{2\pi me}{h^2} \, . \tag{2.14}$$

In the following, we consider $\Psi(\mathbf{r})$ only for $|\mathbf{r}'| \lll |\mathbf{r}|$ (the "asymptotic solution"), so that $\mathbf{r} - \mathbf{r}'$, $\mathbf{k}' := k'\hat{\mathbf{r}}$ and $\mathbf{r}$ are parallel (see Fig. 2.1). Thus the approximation

$$|\mathbf{r} - \mathbf{r}'| = r - \frac{\mathbf{r}' \cdot \mathbf{k}'}{k'} \tag{2.15}$$

is valid and we obtain

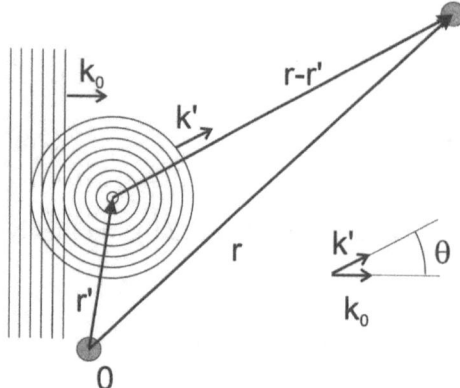

Fig. 2.1. Sketch showing an incident wave with wave vector $\mathbf{k}_0$, scattered by a scattering center at $\mathbf{r}'$, which acts as a source of a secondary spherical wavelet $\Psi_1(\mathbf{r})$

$$\Psi_1(\mathbf{r}) = \sigma \exp(-2\pi i k' r) \int_\Omega \frac{\exp(-2\pi i(\mathbf{k}_0 - \mathbf{k}') \cdot \mathbf{r}')}{|\mathbf{r} - \mathbf{r}'|} \Phi(\mathbf{r}') d^3\mathbf{r}' \ . \tag{2.16}$$

We now introduce the scattering vector $\mathbf{u} = -(\mathbf{k}_0 - \mathbf{k}')$, which is related to the scattering angle θ (see Fig. 2.1) by $u = (2/\lambda)\sin\theta/2$. Replacing $|\mathbf{r} - \mathbf{r}'|$ by r, we obtain:

$$\Psi(\mathbf{r}, \mathbf{u}) = \exp(-2\pi i \mathbf{k}_0 \cdot \mathbf{r}) + \frac{\exp(-2\pi i k' r)}{r} f_e(\mathbf{u}) \ , \tag{2.17}$$

where $f_e(\mathbf{u})$ is the *atomic scattering factor for electrons*, given by

$$\boxed{f_e(\mathbf{u}) = \sigma \int_\Omega \exp(2\pi i \mathbf{u} \cdot \mathbf{r}')\Phi(\mathbf{r}') d^3\mathbf{r}' \ .} \tag{2.18}$$

Note that $f_e(\mathbf{u})$ is also called the *scattering length* because it has dimensions of length. It is proportional to the Fourier transform of $\Phi(\mathbf{r})$, denoted by $\mathcal{F}\Phi(\mathbf{r})$ (see Sect. B.1). Equation (2.18) is analogous to the X-ray scattering factor

$$f_X(\mathbf{u}) = \int_\Omega \exp(2\pi i \mathbf{u} \cdot \mathbf{r}')n(\mathbf{r}') d^3\mathbf{r}' \ , \tag{2.19}$$

where $n(\mathbf{r}')$ is the electron density distribution in the atom. Note that the electron is scattered by the Coulomb potential of the electron shell and the nucleus of the atom, whereas the X-ray scattering is due to the effect of the incident photon forcing an oscillation of the "hit" electron, which thus emits a secondary wave. Poisson's equation provides a relation between the charge density of the electron shell and its contribution to the atomic potential. Values of $f_X(\mathbf{u})$ are available in the literature for most elements. The Mott formula

$$f_e(\mathbf{u}) = \frac{2\pi m e^2}{\epsilon_0 h^2} \frac{(Z - f_X(\mathbf{u}))}{|\mathbf{u}|^2} , \tag{2.20}$$

which is based on Poisson's equation, gives a relation between $f_e(\mathbf{u})$ and $f_X(\mathbf{u})$, where Z is the atomic number of the target atom and ϵ_0 the permittivity of a vacuum. Equation (2.20) has been used to compute $f_e(\mathbf{u})$ [4, 5]. A list of atomic scattering amplitudes $f_e(\mathbf{u})$ is provided in [5], and these are also used in the EMS program package [6]. There, the $f_e(\mathbf{u})$ are given for $m = m_0$ in the parametric form

$$f'_e(s) = \sum_{n=1}^{4} a_n \exp(-b_n s^2) , \text{ where } s = \frac{|\mathbf{u}|}{2} = \frac{\sin(\theta/2)}{\lambda} . \tag{2.21}$$

Here a_n and b_n are the parameters listed in [5], given in units of Å and Å^2, respectively. The relativistically corrected atomic scattering amplitudes are given by

$$f_e(\mathbf{u}) = \sqrt{\left(1 + \frac{h^2}{m_0^2 c^2 \lambda^2}\right)} f'_e\left(\frac{|\mathbf{u}|}{2}\right) . \tag{2.22}$$

Figure 2.2 diplays the atomic scattering amplitudes of Ga, In and As. One can clearly see that $f_e(\mathbf{u})$ depends on the type of scattering atom. This chemical sensitivity is the basis of the technique for compositional analysis described in this book. Figure 2.2 also reveals that the scattering amplitude decreases quickly with increasing scattering parameter $|\mathbf{u}|$.

2.2 Kinematical Approximation

2.2.1 The Structure Amplitude

The scattered electron wave function $\Psi_S(\mathbf{r}, \mathbf{u})$ is given in the first Born approximation (see (2.17)) by

$$\Psi_S(\mathbf{r}, \mathbf{u}) = \frac{\exp(-2\pi i k_0 r)}{r} f_e(\mathbf{u}) . \tag{2.23}$$

We now consider a unit cell of a crystal containing N atoms with atomic scattering factors $f_e^{(i)}$ at positions $\mathbf{r}_i$ $(i = 1, \ldots, N)$ (Fig. 2.3). We substitute $\mathbf{r}' = \mathbf{r}'' + \mathbf{r}_i$ in (2.18) and obtain the *structure amplitude* (also called the

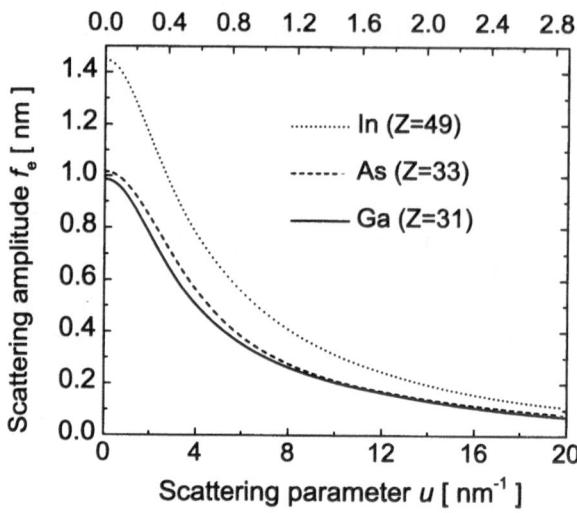

Fig. 2.2. Atomic scattering amplitudes for In, Ga and As plotted versus the scattering parameter $|\mathbf{u}|$, calculated for an accelerating voltage of 200 kV

structure factor) of the unit cell by summing the atomic scattering factors of all atoms in the unit cell according to

$$F_S(\mathbf{u}) = \sum_{i=1}^{N} \sigma \int \Phi^{(i)}(\mathbf{r}'') \exp\{2\pi i \mathbf{u} \cdot (\mathbf{r}'' + \mathbf{r}_i)\} \, d^3\mathbf{r}'' \tag{2.24}$$

$$\Rightarrow \boxed{F_S(\mathbf{u}) = \sum_{i=1}^{N} f_e^{(i)}(\mathbf{u}) \exp\{2\pi i(\mathbf{u} \cdot \mathbf{r}_i)\}} \ . \tag{2.25}$$

2.2.2 The Lattice Amplitude

The interaction of the scattered waves emanating from a periodic assembly of unit cells leads to the effect of electron diffraction. Summing the structure amplitudes of unit cells with their origins at positions $\mathbf{r}_T = n_1\mathbf{r}_1 + n_2\mathbf{r}_2 + n_3\mathbf{r}_3$, where $n_{1,2,3}$ are integers and $\mathbf{r}_{1,2,3}$ the lattice translation vectors, we obtain

$$G(\mathbf{u}) = F_S(\mathbf{u}) \sum_{n_{1,2,3}} \exp\{2\pi i[\mathbf{u} \cdot (n_1\mathbf{r}_1 + n_2\mathbf{r}_2 + n_3\mathbf{r}_3)]\} \quad n_{1,2,3} \in \mathbb{N} \ . \tag{2.26}$$

Figure 2.4 clearly reveals that the modulus $|G(\mathbf{u})|$ is a peaked function, where each peak corresponds to a diffracted beam. Peaks occur for scattering vectors $\mathbf{u}$ that obey

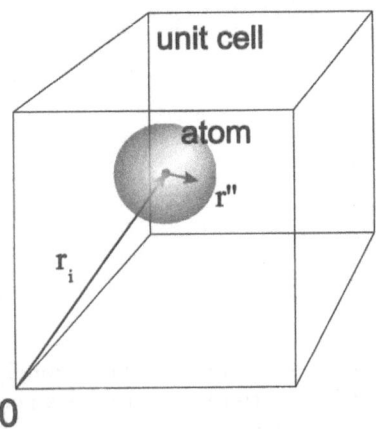

Fig. 2.3. Sketch of a unit cell, with the definition of the vectors $\mathbf{r}_i$ and $\mathbf{r}''$ used in (2.24)

$$\mathbf{u} \cdot (n_1\mathbf{r}_1 + n_2\mathbf{r}_2 + n_3\mathbf{r}_3) = n \quad n \in \mathbb{N} , \tag{2.27}$$

which is fulfilled when $\mathbf{u}$ is a *reciprocal-lattice vector* $\mathbf{g}_{hkl}$ with Miller indices h, k, l, such as

$$\mathbf{g}_{hkl} = h\mathbf{g}_1 + k\mathbf{g}_2 + l\mathbf{g}_3 \quad h, k, l \in \mathbb{N} . \tag{2.28}$$

The reciprocal-lattice base vectors $\mathbf{g}_{1,2,3}$ are defined by

$$\mathbf{g}_i = \frac{\mathbf{r}_j \times \mathbf{r}_k}{\mathbf{r}_i \cdot [\mathbf{r}_j \times \mathbf{r}_k]} , \quad (i, j, k) \in \{(1, 2, 3), (2, 3, 1), (3, 1, 2)\} . \tag{2.29}$$

We thus obtain for the vectors $\mathbf{g}_{hkl}$

$$\mathbf{g}_{hkl} \cdot (n_1\mathbf{r}_1 + n_2\mathbf{r}_2 + n_3\mathbf{r}_3) = hn_1 + kn_2 + ln_3 , \tag{2.30}$$

which proves that $\mathbf{u} = \mathbf{g}_{hkl}$ fulfills (2.27).

2.2.3 The Thin-Foil Specimen

To calculate the electron wave diffracted by a real specimen in the kinematical approximation, we consider a thin foil of thickness t. Summing the contributions from all points of the exit surface of the thin foil, interfering at a point P at a distance $\mathbf{R}$ from the surface as displayed in Fig. 2.5, we obtain, for example from Fresnel's zone construction method [1], the total secondary wave corresponding to the diffraction vector $\mathbf{g}_{hkl}$ as

$$\Psi_S(\mathbf{g}_{hkl}) = \mathrm{i}\lambda t \frac{F_S(\mathbf{g}_{hkl})}{V_C} \exp\{-2\pi\mathrm{i}\mathbf{k}_0 \cdot \mathbf{R}\} , \tag{2.31}$$

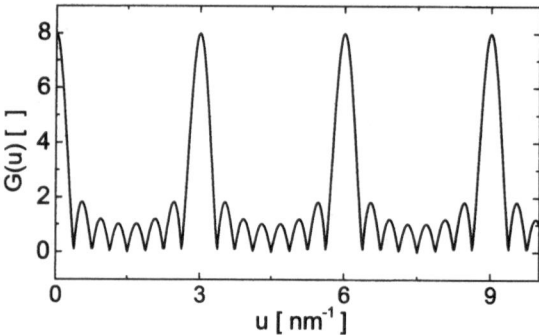

Fig. 2.4. Graph visualizing the effect of diffraction described by the lattice amplitude in (2.26). The function $|G(u)|$ is plotted versus the scattering "vector" u, where $G(u) = \sum_{n=0}^{7} \exp\{2\pi i u(n r_1)\}$, using a 1D lattice translation "vector" $r_1 = (1/3)$ nm. Clearly, the modulus of $G(u)$ contains peaks at positions $u = 0, 3, 6, 9, \ldots$, where $u r_1 = m$, $m \in \mathbb{N}$. The positions **u** of the peaks correspond to the directions where constructive interference takes place, thus describing the propagation directions of diffracted beams

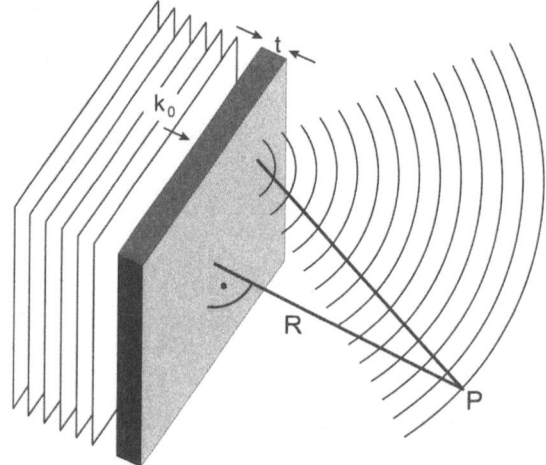

Fig. 2.5. An incident plane wave with wave vector $\mathbf{k_0}$ passing through a thin-foil specimen. At the point P, the wave function is composed of the sum of scattered waves from all points of the surface

where V_C is the volume of the unit cell. In the kinematical approximation, the amplitude of a diffracted beam $\mathbf{g}_{hkl}$ increases linearly with the specimen thickness t. The dependence of the amplitude of the scattered wave upon $1/r$ that is obtained for single-atom scattering in (2.23) has vanished in (2.31). Note that if this approximation were valid, the the amplitude of the diffracted wave would become equal to the amplitude of the incident wave at $t_c = V_C/(F_S(\mathbf{g}_{hkl})\lambda)$ (e.g. $t_c = 15$ nm for $\mathbf{g}_{002}$ in GaAs), whereas the amplitude of the undiffracted beam would not change. This would violate energy conservation, and this consideration makes it clear that the kinematical approximation is valid only for a specimen thickness $t \ll t_c$.

In electron microscopy, the beams that contribute to the image are selected with the objective aperture, which is located in the back focal plane of the objective lens. The intensity of an image that is obtained in such a way that only the beam $\mathbf{g}_{hkl}$ passes through the objective aperture, is given by

$$I(\mathbf{g}_{hkl}) = |\Psi_S(\mathbf{g}_{hkl})|^2 \propto |F_S(\mathbf{g}_{hkl})|^2 . \qquad (2.32)$$

Such a micrograph is called a *single-beam bright-field* (BF) image if $\mathbf{g}_{hkl} = 0$ and a *single-beam dark-field* (DF) image otherwise. If more than one beam $\mathbf{g}_{hkl}$ passes through the objective aperture, the electron waves of all beams interfere in the image plane, and nonlinear imaging theory must be applied to describe the intensity distribution.

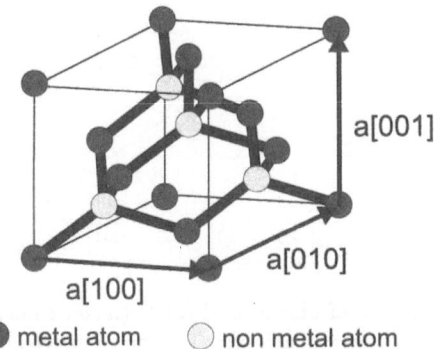

● metal atom   ○ non metal atom

Fig. 2.6. The nonprimitive unit cell of the sphalerite crystal structure. The primitive basis vectors are $\mathbf{r}_1 = (a/2)[110]$, $\mathbf{r}_2 = (a/2)[101]$ and $\mathbf{r}_3 = (a/2)[011]$, where a is the lattice parameter. The basis contains two atoms of different type (metal or nonmetal) at positions $\mathbf{r}_i$ and $\mathbf{r}_i + (1/4)a[111]$. By convention, the origin is occupied by a metal atom. The reciprocal-lattice base vectors are $\mathbf{g}_1 = a^{-1}[\bar{1}\bar{1}1]$, $\mathbf{g}_2 = a^{-1}[\bar{1}1\bar{1}]$ and $\mathbf{g}_3 = a^{-1}[1\bar{1}\bar{1}]$

2.2.4 Chemical Sensitivity

We now use the results of the preceding paragraphs to discuss chemical sensitivity in the kinematical approximation. Equation (2.31) shows that the amplitude of $\Psi_S(\mathbf{g}_{hkl})$ is proportional to $F_S(\mathbf{g}_{hkl})$. We therefore concentrate on the discussion of the structure factor. Since the present book is concerned with composition determination in sphalerite-type crystals (Fig. 2.6), we consider the structure factor of a beam $\mathbf{g}_{hkl}$ for a binary material AC, given by

$$F_S(\mathbf{g}_{hkl}) = 4\left\{f_e^{(A)}(\mathbf{g}_{hkl}) + f_e^{(C)}(\mathbf{g}_{hkl})\exp[2\pi i(h+k+l)/4]\right\}\,. \qquad (2.33)$$

In the case of a ternary material $A_xB_{1-x}C$, we assume a random distribution of two different sorts of metal atoms (e.g. In and Ga for $In_xGa_{1-x}As$) on the metal sublattice. The structure factor of the ternary material is then obtained to a good approximation by assuming, that the atomic scattering amplitude of the metal site is a linear combination of the scattering amplitudes of the two types of metal atoms under consideration:

$$f_e^{(A_xB_{1-x})}(\mathbf{g}_{hkl}) = xf_e^{(A)}(\mathbf{g}_{hkl}) + (1-x)f_e^{(B)}(\mathbf{g}_{hkl})\,. \qquad (2.34)$$

The structure factor of the ternary material is obtained by replacing $f_e^{(A)}(\mathbf{g}_{hkl})$ by $f_e^{(A_xB_{1-x})}(\mathbf{g}_{hkl})$ in (2.33).

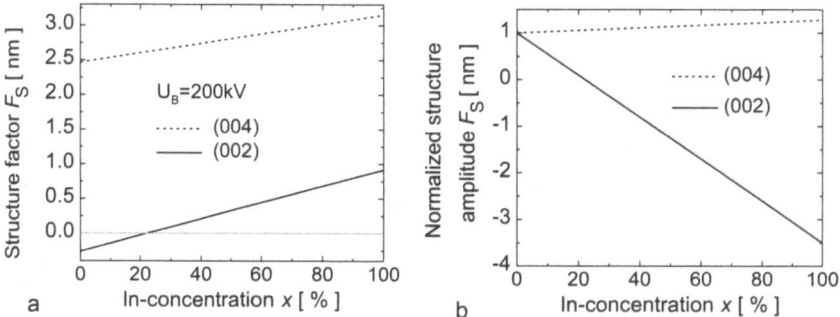

Fig. 2.7. (a) Structure factors of the (002) (*solid curve*) and (004) (*dashed curve*) beams in $In_xGa_{1-x}As$, plotted versus the In concentration x. The curves were computed assuming an acceleration voltage of 200 kV. (b) The normalized structure factors $F_S(\mathbf{g}_{hkl}, x)/F_S(\mathbf{g}_{hkl}, 0)$, which determine the chemical sensitivity as indicated in 2.36

In the case of a sphalerite-type material, the {002} beam is called *chemically sensitive*. Its structure factor is given by

$$F_S(\mathbf{g}_{002}) = 4\left\{f_e^{(A_xB_{1-x})}(\mathbf{g}_{002}) - f_e^{(C)}(\mathbf{g}_{002})\right\}\,. \qquad (2.35)$$

Figure 2.7a shows the structure factors of the (002) and (004) beams (which will be important in relation to the CELFA technique) in $In_xGa_{1-x}As$ plotted versus the In concentration x. Figure 2.7a reveals that the structure amplitude of the {002} beam is smaller than that of the {004} beam. An important factor for composition analysis in practice is the *relative* change of the amplitude as a function of the composition, given by

$$S(\mathbf{g}_{hkl}) = \frac{\partial}{\partial x}\left(\frac{F_S(\mathbf{g}_{hkl}, x)}{F_S(\mathbf{g}_{hkl}, 0)}\right)\,, \qquad (2.36)$$

which we define as the *chemical sensitivity*. Figure 2.7b shows the normalized structure factor $F_S(\mathbf{g}_{hkl}, x)/F_S(\mathbf{g}_{hkl}, 0)$ plotted versus the In concentration. Its slope clearly is larger for (002) than for (004), revealing the higher chemical sensitivity of the (002) beam.

Note that the slope of the *normalized* structure factor of a beam is defined as the chemical sensitivity in (2.36), and not the slope of the structure factor itself, which is displayed in Fig. 2.7a and is similar for the (002) and (004) beams. The definition in (2.36) was chosen for the purpose of composition determination in real specimens, as discussed in the following.

Theoretically, one can assume a perfect specimen that has a constant thickness, is perfectly clean and has a spatially constant orientation relative to the incident electron beam, investigated in an ideal microscope that is perfectly stable so that infinitely long exposure times can be attained, with "perfect" electrons that interact only elastically. In that case, all beams with an amplitude that depends on the concentration in some unambiguous manner could be used for composition determination. However, in practice, neither the specimen, nor the microscope is perfect. The thickness of the specimen changes on a nanometer scale in an HRTEM image. The specimen surface is covered by an amorphous layer due to the preparation of the TEM specimen unless the specimen was prepared by cleavage. Even then, thin oxide layers can be found on the surfaces. The lattice parameter of the layer under investigation differs in most cases from that of the substrate, and the resulting strain combined with thickness fluctuations results in a local bending of the thin-foil specimen. Thus, the orientation relative the electron beam varies slightly. Also, the microscope is not ideal. In addition to lens aberrations, its limited stability allows exposure times of at most a few seconds. Since one leaves wave optics behind in the image plane, where particle statistic becomes important, the short exposure time adds to the noise in the image. Inelastically scattered electrons give images different from that produced by elastically scattered electrons, mainly because the focal length of the objective lens depends on the wavelength of the electron beam. All these effects lead to an image that shows a local variation of the beam amplitudes due to thickness fluctuations and lattice bending and contains noise due to electron statistics and amorphous surface layers. How is this related to chemical sensitivity? Consider the following example. A thickness fluctuation of Δt leads (in the kinematical approximation) to a change of the relative amplitude $|\Psi_S(g_{hkl}, t + \Delta t)|/|\Psi_S(g_{hkl}, t)|$ of the beam $\mathbf{g}_{hkl}$ equal to $\Delta t/t$, as is obvious from (2.31). For the (004) reflection, a thickness fluctuation of 10% has the same effect as a composition fluctuation of 40% as can be deduced from Fig. 2.7b. For (002), the corresponding fluctuation of the composition is only 4%. This example clearly shows that the maximum *relative* change of the structure amplitude with composition is desirable, corresponding to the maximum chemical sensitivity as defined in (2.36).

It was mentioned above that the kinematical approximation of electron diffraction is valid only for a small specimen thickness, well below ≈ 10 nm. To obtain beam amplitudes valid for realistic thicknesses of 10 to 20 nm, the dynamical diffraction theory has to be applied. In the next section, we give a brief outline of the Bloch wave theory, which provides the data about the amplitudes and phases of beams used in composition evaluation by the CELFA technique.

2.3 The Bloch Wave Approach

The starting point is again the Schrödinger equation (2.1), but $\Phi(\mathbf{r})$ now describes the crystal potential. $\Phi(\mathbf{r})$ is expanded as a Fourier series based on the reciprocal lattice:

$$\Phi(\mathbf{r}) = \sum_{\mathbf{g} \neq 0} \frac{F_S(\mathbf{g})}{\Omega_C} \exp(-2\pi i \mathbf{g} \cdot \mathbf{r}) + \frac{F_S(0)}{\Omega_C} \,, \tag{2.37}$$

where Ω_C is the volume of the crystal unit cell. Using a Bloch wave ansatz

$$\Psi(\mathbf{r}) = u_{\mathbf{k}}(\mathbf{r}) \exp(-2\pi i \mathbf{k} \cdot \mathbf{r}) \,, \tag{2.38}$$

where $u_{\mathbf{k}}(\mathbf{r})$ has the periodicity of the lattice, i.e.

$$u_{\mathbf{k}}(\mathbf{r}) = \sum_{\mathbf{g}} C_{\mathbf{g}} \exp(-2\pi i \mathbf{g} \cdot \mathbf{r}) \tag{2.39}$$

(the dependence of the $C_{\mathbf{g}}$ on $\mathbf{k}$ is not explicitly indicated here), we obtain

$$\Psi(\mathbf{r}) = \sum_{\mathbf{g}} C_{\mathbf{g}} \exp(-2\pi i (\mathbf{g} + \mathbf{k}) \cdot \mathbf{r}) \,, \tag{2.40}$$

where the $C_{\mathbf{g}}$ are called the *Bloch wave coefficients*. The constants may be collected together as follows:

$$V_{\mathbf{g}} = \frac{2me}{h^2} \frac{F_S(\mathbf{g})}{\Omega_C} \,,$$

$$\kappa^2 = \frac{2me}{h^2} (E + V_0) \,, \tag{2.41}$$

where κ is the mean electron wavenumber in the crystal and V_0 is the mean inner potential. Inserting (2.37), (2.40) and (2.41) into (2.1) we obtain

$$\sum_{\mathbf{g}} C_{\mathbf{g}} \exp(-2\pi i (\mathbf{g} + \mathbf{k}) \cdot \mathbf{r}) \left[\kappa^2 - (\mathbf{g} + \mathbf{k})^2 \right]$$

$$+ \sum_{\mathbf{h}} \sum_{\mathbf{g}' \neq 0} V_{\mathbf{g}'} C_{\mathbf{h}} \exp(-2\pi i (\mathbf{h} + \mathbf{k} + \mathbf{g}') \cdot \mathbf{r}) = 0 \,. \tag{2.42}$$

Substituting $\mathbf{h} + \mathbf{g}' = \mathbf{g}$, the double sum becomes

$$\sum_{\mathbf{h}}\sum_{\mathbf{g}\neq\mathbf{h}} V_{\mathbf{g}-\mathbf{h}}C_{\mathbf{h}}\exp(-2\pi i(\mathbf{g}+\mathbf{k})\cdot\mathbf{r}) = \sum_{\mathbf{h}\neq\mathbf{g}}\sum_{\mathbf{g}}\cdots \quad . \tag{2.43}$$

Equation (2.42) is fulfilled for all $\mathbf{r}$ if

$$\forall \mathbf{g}: \quad \left[\kappa^2 - (\mathbf{g}+\mathbf{k})^2\right]C_{\mathbf{g}} + \sum_{\mathbf{h}\neq\mathbf{g}} V_{\mathbf{g}-\mathbf{h}}C_{\mathbf{h}} = 0 \ . \tag{2.44}$$

Equation (2.44) is called the *secular equation*. We now restrict ourselves to beams $\mathbf{g}$ in the zero-order Laue zone (ZOLZ), which is a good approximation close to a zone axis (ZA) orientation in high-energy electron diffraction because the scattering amplitude decreases quickly with increasing scattering angle (see Figs. 2.2 and 2.8).

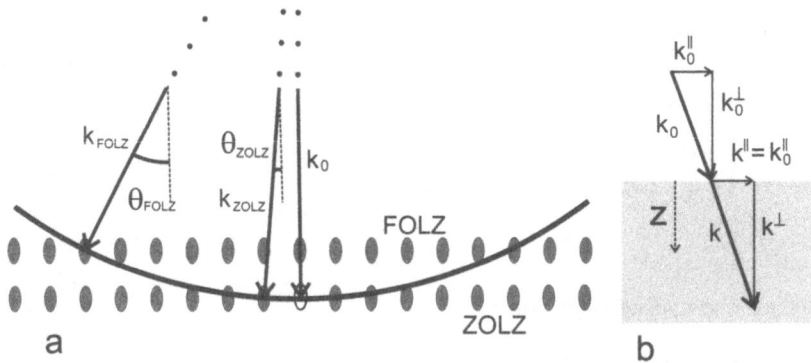

Fig. 2.8. (a) Sketch showing the reciprocal lattice and the Ewald sphere. The row of reciprocal-lattice points that contains the origin is called the zero-order Laue zone (ZOLZ). The next row in the direction $-\mathbf{k}^0$ is the first-order Laue zone (FOLZ). The scattering angle of a diffracted beam is larger for the FOLZ than for the ZOLZ. (b) defines the z direction and visualizes the decomposition of the wave vectors $\mathbf{k}_0$ of the incident beam and $\mathbf{k}$ of a Bloch wave in the specimen into components $\mathbf{k}_{(0)}^{\parallel}$ and $\mathbf{k}_{(0)}^{\perp}$ tangential and perpendicular, respectively, to the specimen surface. The continuity of the electron wave function at the substrate surface yields $\mathbf{k}_0^{\parallel} = \mathbf{k}^{\parallel}$

For convenience, we define the z direction as the downward normal to the crystal surface. The surface is assumed to be parallel to the $\mathbf{g} \in$ ZOLZ (see Fig. 2.8). The wave vectors $\mathbf{k}_0$ of the incident beam and $\mathbf{k}$ of the Bloch wave in the specimen are expanded into components $\mathbf{k}_{(0)}^{\parallel}$ and $\mathbf{k}_{(0)}^{\perp}$ parallel and perpendicular, respectively, to the specimen surface. It follows immediately from the continuity of the wave function at the "upper" surface of the specimen ($z = 0$) that $\mathbf{k}_0^{\parallel} = \mathbf{k}^{\parallel}$. We obtain for the expression in square brackets in (2.44)

$$\kappa^2 - (\mathbf{k} + \mathbf{g})^2 = \kappa^2 - (\mathbf{k}^\| + \mathbf{k}^\perp + \mathbf{g})^2 = (\kappa^2 - \mathbf{k}^{\perp^2}) - (\mathbf{k}^\| + \mathbf{g})^2 , \quad (2.45)$$

since $\mathbf{k}^\| \cdot \mathbf{k}^\perp = \mathbf{k}^\perp \cdot \mathbf{g} = 0$. We now introduce the high-energy approximation $\kappa \approx \mathbf{k}^\perp$. If we understand $\kappa^2 - \mathbf{k}^{\perp^2}$ as a function of $k^\perp$, a first-order Taylor expansion yields

$$\kappa^2 - \mathbf{k}^{\perp^2} \approx 2\kappa(\kappa - k^\perp) . \quad (2.46)$$

Using the high-energy approximation in (2.44), we obtain

$$-(\mathbf{k}^\| + \mathbf{g})^2 C_\mathbf{g} + \sum_{\mathbf{h} \neq \mathbf{g}} V_{\mathbf{g}-\mathbf{h}} C_\mathbf{h} = 2\kappa(k^\perp - \kappa)C_\mathbf{g} . \quad (2.47)$$

For convenience, we now change the notation for the reciprocal-lattice vectors of the ZOLZ by denoting them by $\mathbf{g}_n$, $n = 1, \ldots, N$. The exact solution of the Schrödinger equation would require $N \to \infty$. In practice, $N \approx 300$ is used for Bloch wave calculations close to a ZA orientation and $N \approx 15$ for a specimen orientation a few degrees off the ZA. Defining the matrix $\mathbf{D}$ and the column vector $\mathbf{C}$ as

$$[\mathbf{D}]_{mn} := \begin{cases} -(\mathbf{k}^\| + \mathbf{g}_m)^2 & \text{for } m = n \\ V_{\mathbf{g}_m - \mathbf{g}_n} & \text{otherwise} \end{cases} , \quad (2.48)$$

$$[\mathbf{C}]_n := C_{\mathbf{g}_n} ,$$

we obtain an eigenvalue equation:

$$\mathbf{D}\mathbf{C} = 2\kappa(k^\perp - \kappa)\mathbf{C} . \quad (2.49)$$

The matrix $\mathbf{D}$ is Hermitian because $V_\mathbf{g} = V_{-\mathbf{g}}^*$. Note that only elastic diffraction, without absorption, is assumed here. If we were to take account of absorption, the matrix $\mathbf{D}$ would be a complex general matrix [3]. Equation (2.49) has N eigenvalues $2\kappa(k^{\perp (j)} - \kappa)$, $j = 1, \ldots, N$, and N eigenvectors $\mathbf{C}^{(j)}$, which are assumed to be normalized. Note that $k^\|$ is constant and is given by the quantum-mechanical continuity of Ψ at the crystal surface. The eigenvector matrix

$$[\mathbf{C}]_{jn} := C_{\mathbf{g}_n}^{(j)} , \quad n, j = 1, \ldots, N, \quad (2.50)$$

is unitary. The total wave function, according to (2.40), is given by

$$\Psi(\mathbf{r}) = \sum_{j=1}^{N} \alpha^{(j)} \Psi^{(j)}(\mathbf{r}) , \quad (2.51)$$

$$\Psi^{(j)}(\mathbf{r}) = \sum_{n=1}^{N} C_{\mathbf{g}_n}^{(j)} \exp\left[-2\pi i \left(\mathbf{k}^{\perp (j)} + \mathbf{k}^\| + \mathbf{g}_n \right) \cdot \mathbf{r} \right] .$$

The $\alpha^{(j)}$ are called the *excitation amplitudes* of the jth Bloch wave $\Psi^{(j)}(\mathbf{r})$. They are obtained from the continuity of Ψ at $z = 0$, where $C_{\mathbf{g}}^{(j)} = 0 \quad \forall \mathbf{g} \neq 0$. We thus obtain the following from (2.51):

$$\Psi_{z=0}(\mathbf{r}) = \sum_{j=1}^{N} \alpha^{(j)} C_0^{(j)} \exp\left[-2\pi i(\mathbf{k}_\| \cdot \mathbf{r})\right] . \tag{2.52}$$

The amplitude of $\Psi_{z=0}(\mathbf{r})$ must be equal to that of the incident electron wave, and thus

$$\sum_{j=1}^{N} \alpha^{(j)} C_0^{(j)} = 1 \quad \Leftrightarrow \quad \alpha^{(j)} = C_0^{\star (j)} . \tag{2.53}$$

Finally, the diffracted beam $\Psi_{\mathbf{g}_n}$ is given by:

$$\Psi_{\mathbf{g}_n} = \sum_{j=1}^{N} C_0^{\star (j)} C_{\mathbf{g}_n}^{(j)} \exp\left[-2\pi i \left(\mathbf{k}^{\perp (j)} + \mathbf{k}^\| + \mathbf{g}_n\right) \cdot \mathbf{r}\right] . \tag{2.54}$$

For use in the following sections, we calculate the diffracted beam at the "lower" surface of q specimen of thickness t. We decompose $\mathbf{r}$ into decomposed into components $\mathbf{r}^\|$ and $\mathbf{r}^\perp = t\hat{z}$ parallel and perpendicular to the specimen surface. We define

$$F(\mathbf{g}_n) := \sum_{j=1}^{N} C_0^{\star (j)} C_{\mathbf{g}_n}^{(j)} \exp\left(-2\pi i k^{\perp (j)} t\right) \tag{2.55}$$

and obtain the final result,

$$\boxed{\Psi_{\mathbf{g}_n}(\mathbf{r}) = \exp\left\{-2\pi i (\mathbf{g}_n + \mathbf{k}_0^\|) \cdot \mathbf{r}^\|\right\} F(\mathbf{g}_n) .} \tag{2.56}$$

$F(\mathbf{g}_n)$ is called the *amplitude of the beam* $\mathbf{g}_n$. The space vector $\mathbf{r}^\|$ lies in the image plane. The wave vector $\mathbf{k}_0^\|$ arises from the tilt of the incident electron beam relative to the normal to the specimen surface (Fig. 2.8). The tilt causes a shift of the diffraction pattern in the back focal plane of the objective lens. The beam amplitudes $F(\mathbf{g}_n)$ will be used extensively in the following sections. Note that they are complex-valued quantities.

Figure 2.9 shows an example of the dependence of the beam amplitudes on the specimen thickness in the electron beam direction as obtained from a Bloch wave computation with the EMS program package [6]. Figure 2.9a shows the amplitude of the (002) beam for two different orientations of the incident electron beam with respect to the specimen. The gray curve corresponds to an exact [100] ZA orientation, with the electron beam parallel to the [100] direction of the specimen. The black curve displays the result for

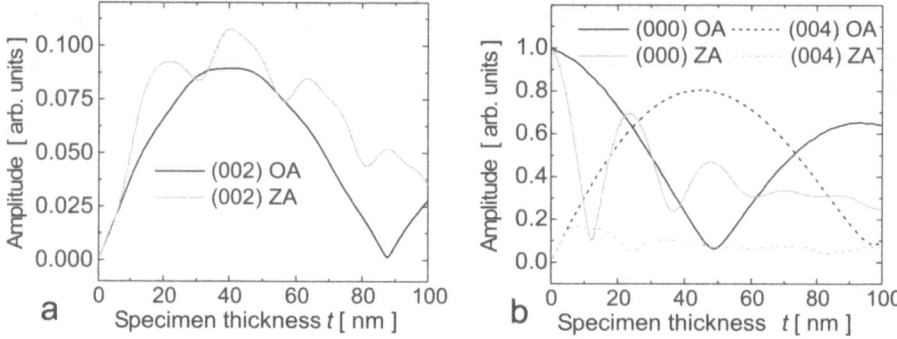

Fig. 2.9. (a) Amplitude of the (002) beam plotted versus the specimen thickness, calculated with the Bloch wave method for GaAs. The *gray line* corresponds to a ZA orientation of the specimen, with the incident electron beam along the [100] direction. The *black line* was computed for an off-ZA (OA) orientation, with the electron beam tilted 5° around the [001] direction and the (004) beam strongly excited, corresponding to the center of the Laue circle (for an explanation see Fig. 2.10) in the position [0 20 2]. (b) Amplitudes of the (000) and (004) beams, presented analogously to (a)

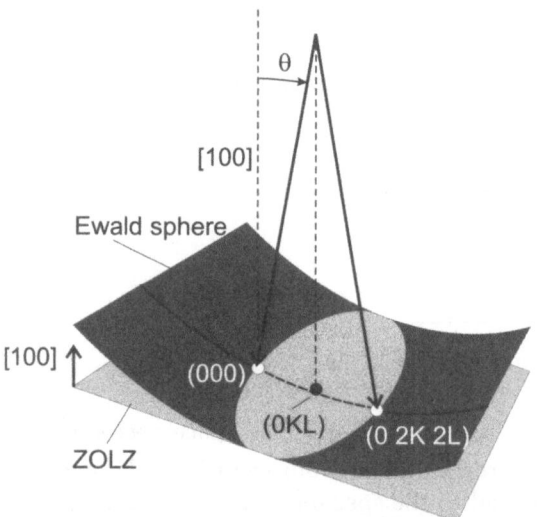

Fig. 2.10. Schematic drawing explaining the meaning of the center of the Laue circle. The incident electron beam is tilted by an angle θ around the [100] ZA. The intersection of the Ewald sphere with the plane of the ZOLZ forms a circle with its center corresponding to the position $(0KL)$ of reciprocal space, called the *center of the Laue circle*. All beams $(0, k, l)$ that fulfill $(k - K)^2 + (l - L)^2 = K^2 + L^2$ lie on the Laue circle and are thus strongly excited

an electron beam direction tilted 5^{circ} from the exact [100] ZA orientation. The *center of the Laue circle* (illustrated in Fig. 2.10) lies at the position $(0,\ K = 20,\ L = 2)$. In this case the beam $(0,\ k = 0,\ l = 4)$ is strongly excited, because it lies on the Laue circle, as becomes obvious from the condition $(k-K)^2 + (l-L)^2 = K^2 + L^2$ given in the caption of Fig. 2.10. Figure 2.9a shows clearly that the off-ZA orientation leads to a much smoother curve for the amplitude of the (002) beam (black curve) compared with the exact ZA orientation (gray curve). The amplitudes of the (000) and (004) beams show a similar behavior (Fig. 2.9b). The complex relationship between the beam amplitude and the specimen thickness that prevails at the exact ZA orientation results from the complicated interaction of many contributing beams. The underlying reason is the short wavelength of the electron beam compared with the crystal lattice parameter, which results in a small curvature of the Ewald sphere. As a consequence, many beams are close to the Ewald sphere and are thus strongly excited in the exact ZA orientation. An electron beam inclined with respect to the ZA leads to an increased distance of the reciprocal-lattice points from the Ewald sphere for most beams, which are therefore only weakly excited. The above considerations clearly show that an off-ZA orientation of the specimen is better suited for a compositional analysis that is based on the evaluation of the amplitude of the (002) beam. The relationship between amplitude and specimen thickness is simpler, and small fluctuations of the specimen thickness exert less influence on the results of the evaluation.

2.4 Channeling Theory

This section sets out the theoretical basis for the technique of strain state analysis, where the positions of local extrema of intensity of either HRTEM images or reconstructed wave functions are used to measure the local lattice parameter or displacement. In the case of a ZA orientation of the specimen, this approach is based on the assumption that the positions of local maxima (or minima) of the intensity describe the positions of atomic columns in the specimen. Note that a ZA orientation, with many beams contributing to image formation, is not necessary if one is interested in the local spacings of one set of lattice planes only. An example is an investigation of pseudomorphically grown buried layers, where the use of lattice planes parallel to the interfaces would suffice for obtaining the local composition. In this case, a two-beam condition would be chosen and the resulting lattice fringe pattern could be interpreted exactly by means of Bloch wave theory. On the other hand, an investigation of the strain state of free-standing islands, where we would be interested in the lattice parameters both parallel and perpendicular to the substrate surface, would require the observation of at least two different sets of lattice planes. In this case, a ZA orientation of the specimen is required and a "dot" pattern is obtained under most imaging conditions.

The positions of the intensity maxima (or minima) are in this case expected to represent the atomic columns of the sample. Even assuming ideal imaging conditions, this effect is not intuitively understandable using a many-beam description in reciprocal space.

On the other hand, it is possible to give at least a qualitative explanation by means of channeling theory, which therefore is outlined here. The description of channeling theory presented in [7] starts with the time-dependent Schrödinger equation

$$-\frac{\hbar}{i}\frac{\partial}{\partial t}\Psi(\mathbf{r}^{\|},t) = \mathbb{H}\Psi(\mathbf{r}^{\|},t) , \qquad (2.57)$$

where

$$\mathbb{H} = -\frac{\hbar^2}{2m}\Delta_{\mathbf{r}^{\|}} - e\Phi(\mathbf{r}^{\|}) ; \qquad (2.58)$$

here $\Delta_{\mathbf{r}^{\|}}$ is the Laplacian operator acting in the $(\mathbf{r}^{\|})$ plane, perpendicular to the electron beam direction z. Equation (2.57) thus describes a (two-dimensional) plane electron wave under the influence of a potential $\Phi(\mathbf{r}^{\|})$ that is averaged along the electron beam direction. The solution of (2.57), expanded in eigenfunctions of the Hamiltonian, is given by

$$\Psi(\mathbf{r}^{\|},z) = \sum_n C_n\varphi(\mathbf{r}^{\|})\exp\left(-i\pi\frac{E_n}{E_0}kz\right) , \qquad (2.59)$$

where $\mathbb{H}\varphi_n = E_n\varphi_n$, and $E_0 = h^2k^2/(2m)$ is the energy of the incident electron wave, with wave number k. The description of dynamical electron diffraction in the framework of channeling theory presented in [7] replaces the z-dependent atomic potential by the projected potential $\Phi(\mathbf{r}^{\|})$ describing the potential of an atomic column. Since the projected two-dimensional column potential has only a very few deep energy states with $E_n \gtrsim E_0/(kz)$, only a few levels E_n contribute to the sum in (2.59). If the overlap between adjacent columns is small, only the radially symmetric states are excited. In [7] it is pointed out that only one state $E_n = E_i$ appears, which can be compared with the 1S state of an atom. Using the boundary condition

$$\sum_n C_n\varphi(\mathbf{r}^{\|}) = \Psi(\mathbf{r}^{\|},z=0) = 1 \qquad (2.60)$$

and the assumptions outlined above, the authors of [7] finally obtain the wave function for *one isolated* atomic column centered at $\mathbf{r}^{\|} = 0$:

$$\boxed{\Psi(\mathbf{r}^{\|},z) - 1 = 2\varphi(\mathbf{r}^{\|})\sin\left\{\frac{\pi E_i}{2E_0}kz\right\}\exp\left\{-i\left[\frac{\pi}{2} + \frac{\pi}{2}\frac{E_i}{E_0 kz}\right]\right\}} . \qquad (2.61)$$

Equation (2.61) reveals an oscillation of the amplitude of $\Psi(\mathbf{r}^{\|},z)$ that depends on the penetration depth (Fig. 2.11b) and a radial amplitude distribution given by the 1S eigenstate $\varphi(\mathbf{r}^{\|})$ (Fig. 2.11a). The authors of [7] showed that the description presented in (2.61) also holds for an atomic column in a crystal where each column is surrounded by neighboring columns. These authors found that $E_i \propto Z/d^{5/4}$, where d is the spacing of the atoms inside one column and Z is the atomic number. They found further that the scaled shape of the eigenstate $\varphi(\mathbf{r}^{\parallel})$ shown in Fig. 2.11a is very similar for different atoms. Empirically, its width seems to scale empirically with $\sqrt{E_i}$.

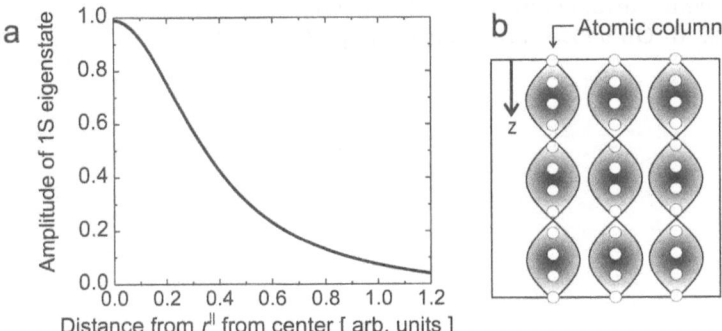

Fig. 2.11. (**a**) Scaled amplitude of the 1S eigenstate $\varphi_i(\mathbf{r}^{\parallel})$ calculated from the channeling eigenvalue problem, plotted versus the scaled distance $\mathbf{r}^{\parallel}$ from the center of the atomic column, taken from [7]. (**b**) visualizes the result of channeling theory given in (2.61). The amplitude of the 1S eigenstate $\varphi(\mathbf{r}^{\parallel})$ oscillates as a function of the penetration depth z. Laterally, the amplitude is peaked at the centers of the atomic columns. The amplitude decreases with increasing distance from the centers of the columns, as shown in (a)

In the context of the strain state analysis, the most interesting feature of (2.61) is the decrease of the amplitude of $\Psi(\mathbf{r}^{\parallel}, z)$ with increasing distance $\mathbf{r}^{\parallel}$ from the center of the column with a slope given by $\varphi(\mathbf{r}^{\parallel})$, the 1S eigenfunction, shown in Fig. 2.11a. This equation thus reveals that the local intensity maxima of the electron wave function at the object exit plane yield the positions of atomic columns. As pointed out in [7], the relationship between the extrema of the intensity and the positions of atomic columns also holds for an HRTEM image obtained with an objective lens that has only rotationally symmetric aberrations. On the other hand, (2.61) is only qualitative, because it takes into account only one eigenstate of the channeling eigenvalue problem. Furthermore, it is only valid for an exact ZA orientation, which clearly is not obtained in practice.

References

1. H. Alexander: *Physikalische Grundlagen der Elektronenmikroskopie*, 1st edn. (Teubner, Stuttgart, 1997)

2. C.J. Humphreys, E.G. Bithell: "Electron diffraction theory". In: *Electron Diffraction Techniques.* ed. by J.M. Cowley (Oxford University Press, Oxford, 1992), p. 75
3. J.M. Cowley: "Electron diffraction: an introduction". In: *Electron Diffraction Techniques.* ed. by J.M. Cowley (Oxford University Press, Oxford, 1992), p. 152
4. G.H. Smith, R.E. Burge: Acta Cryst. **15**, 15 (1962)
5. P.A. Doyle, P.S. Turner: Acta Cryst. A **24**, 390 (1968)
6. P.A. Stadelmann: Ultramicroscopy **51**, 131 (1987)
7. D. Van Dyck, M. Op de Beeck: Ultramicroscopy **64**, 99 (1996)

3 Image Formation

The starting point for this chapter is the wave function of the electron at the object exit surface. Here we follow the electron wave from the object to the image and describe the nonlinear process of image formation. We start with the imaging process assuming an ideal microscope. The description is then refined by taking into account wave aberrations induced by the objective lens. In the next step, we consider effects of incoherence caused by fluctuations of the lens currents and of the high tension, and by electron energy loss. The interpretation of TEM images is further complicated by the loss of the phase of the electron wave because only the image intensity can be recorded in the image plane. Two methods will be briefly outlined that may be able to (approximately) solve this inverse problem of phase retrieval, namely off-axis electron holography and focal-variation reconstruction. For further information, see [1, 2, 3, 4, 5, 6].

3.1 Nonlinear Coherent Imaging

In an electron microscope, the wave function at the object exit surface is magnified in several steps. The first lens is the objective lens. Its magnification lies between 20 and 100 times. If we follow the path of a single ray through the individual magnifying lenses, we find that its angle with respect to the optic axis (*angle of incidence*) is inversely proportional to the image magnification. The next lens after the objective lens is the intermediate lens. There, the angle of incidence of the rays has been decreased by one to two orders of magnitude. Because the lens aberrations decrease strongly with decreasing angle of incidence, the aberrations of all lenses following the objective lens can be neglected. Therefore, only the objective lens is important for image formation.

Figure 3.1 shows the electron beam passing through the specimen and the objective lens. The incident electron beam is diffracted in the crystalline specimen. The electron wave propagates towards the objective lens, which generates a diffraction pattern in the back focal plane; this pattern can be described as the Fourier transform $\mathcal{F}\Psi(\mathbf{r})$ of the wave function at the object exit plane $\Psi(\mathbf{r})$. The space vector $\mathbf{r}$ is from now on assumed to be two-dimensional and to lie in a plane perpendicular to the electron beam

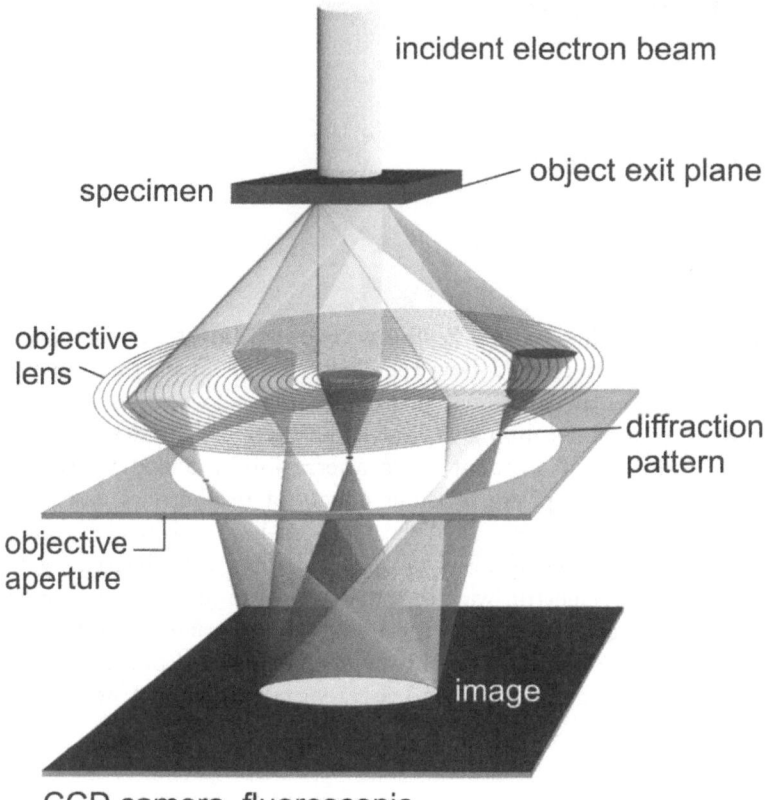

incident electron beam

object exit plane

specimen

objective
lens

diffraction
pattern

objective
aperture

image

CCD camera, fluoroscopic

Fig. 3.1. Schematic drawing showing the functional principle of the transmission electron microscope. Each diffracted beam is focused to a spot in the diffraction pattern, which occurs in the back focal plane of the objective lens. Beams selected with the objective aperture interfere in the image [Please see Plate 1 for color reproduction of this plate]

direction. The position and size of the objective aperture in the back focal plane determine the beams that contribute to the image. The propagation of the electron wave from the diffraction plane to the image corresponds to an inverse Fourier transform. Assuming an "ideal" objective lens, the image intensity $|\mathcal{F}^{-1}\mathcal{F}\Psi(\mathbf{r})|^2 = |\Psi(\mathbf{r})|^2$ is given by the intensity of the (magnified) wave function at the object exit surface $\Psi(\mathbf{r})$.

3.1.1 Influence of Aberrations and Defocus of the Objective Lens

In reality, lens aberrations lead to a distortion of the electron wave. The most prominent type of aberration is the spherical aberration of the objective lens. Its effect on the imaging is described by a phase shift of the electron wave in the back focal plane of the objective lens given by

$$\boxed{\chi_S(\mathbf{k}) = \frac{1}{2}\pi C_S \lambda^3 (\mathbf{k} \cdot \mathbf{k})^2 \, ,}$$

(3.1)

where $\mathbf{k}$ is the coordinate of a point in the diffraction pattern in reciprocal space and C_S is the *spherical-aberration constant*. Note that a spherical-aberration-corrected transmission electron microscope has recently been developed [7]. The experimental investigations that will be described in this book were carried out mainly with two different transmission electron microscopes. The parameters of those microscopes are listed in Table 3.1.

Table 3.1. Parameters of TEMs frequently used in the work described in this book. U_{max} is the maximum available value of the accelerating voltage, C_S is the spherical-aberration constant and Δ is the mean focal spread

Type of TEM	U_{max}	C_S	Δ
Philips CM200/ST FEG	200 kV	1.2 mm	6.8 nm
Philips CM30/UT FEG	300 kV	0.65 mm	6.8 nm

The second type of image distortion is caused by inexact focusing of the objective lens. For an exactly focused specimen, the relation between the object plane (i.e. the lower surface of the specimen) and the image plane of a lens of *focal length* f can be described using the distances from the center of the lens to the object plane d_{OP} and to the image plane d_{IP} by the following relation:

$$\frac{1}{f} = \frac{1}{d_{OP}} + \frac{1}{d_{IP}} \, .$$

(3.2)

The focal length of the objective lens is adjusted via its current. In the TEM, the distance d_{IP} is fixed and d_{OP} is defined by the position of the specimen; $d_{IP} \gg d_{OP}$. A deviation Δf from the focal length f defined by (3.2) leads to a shift Δd_{OP} of the image plane that is in focus, according to

$$\frac{1}{f + \Delta f} = \frac{1}{d_{OP} + \Delta d_{OP}} + \frac{1}{d_{IP}} \approx \frac{1}{d_{OP} + \Delta d_{OP}} \Rightarrow \Delta d_{OP} \approx \Delta f \, .$$

(3.3)

The important quantity Δf is called the *defocus* of the objective lens. According to (3.3), the defocus is negative (known as *underfocus*) if the plane that is in focus is shifted in the direction of the electron beam from the "lower" specimen surface. A positive defocus value is called an *overfocus*. The influence of defocus on the imaging can also be described by a phase shift of the electron wave in the back focal plane of the objective lens, given by

$$\boxed{\chi_{\Delta f} = \pi\,\Delta f\,\lambda(\mathbf{k}\cdot\mathbf{k})\,.}$$
(3.4)

Neglecting effects of incoherence, the image intensity now reads:

$$I(\mathbf{r}) = \mathcal{F}^{-1}\left\{\exp\left[i(\chi_{\mathrm{S}}(\mathbf{k}) + \chi_{\Delta f}(\mathbf{k}))\right]\,\mathcal{F}\Psi(\mathbf{r})\right\}\,.$$
(3.5)

3.1.2 Nonlinear Imaging Theory

Composition evaluation by the CELFA technique exploits the image *diffractogram*, defined by

$$\widetilde{I}(\mathbf{k}) = \mathcal{F}I(\mathbf{r}),$$
(3.6)

where $I(\mathbf{r})$ is the image intensity. Assuming a crystalline specimen, the image intensity shows a periodic pattern if more than one beam passes through the objective aperture. Thus, the Fourier transform of the image contains *reflections* related to the diffraction pattern. However, one has to bear in mind that the diffractogram is different from the diffraction pattern even for an ideal objective lens, because $\mathcal{F}|\Psi(\mathbf{r})|^2 \neq \mathcal{F}\Psi(\mathbf{r})$. In this section, the relationship between the reflections contained in the diffractogram and the beams in the diffraction pattern is outlined.

Presupposing a crystalline specimen, the wave function at the object exit surface is described by a sum over all beams $\Psi_{\mathbf{g}_n}(\mathbf{r})$, $n = 1,\ldots,N$, given by (2.56):

$$\Psi(\mathbf{r}) = \sum_{n=1}^{N}\exp\left(-2\pi i\mathbf{g}'_n\cdot\mathbf{r}\right)F(\mathbf{g}_n)\,,$$
(3.7)

where we have used the definition

$$\mathbf{g}'_n := \mathbf{g}_n + \mathbf{k}_0^{\parallel}\,.$$
(3.8)

The diffraction pattern is the Fourier transform of $\Psi(\mathbf{r})$ as follows:

$$\widetilde{\Psi}(\mathbf{k}) = \sum_{n=1}^{N}F(\mathbf{g}_n)\delta(\mathbf{k} - \mathbf{g}'_n)$$
(3.9)

(see (B.13)). Taking into account lens aberrations, each of the beams gains a phase shift

$$t(\mathbf{k}) = \exp\left\{i\chi(\mathbf{k})\right\}\,,\quad \chi(\mathbf{k}) := \chi_{\mathrm{S}}(\mathbf{k}) + \chi_{\Delta f}(\mathbf{k})\,,$$
(3.10)

where the wave aberrations $\chi_{\mathrm{S}}(\mathbf{k})$ and $\chi_{\Delta f}(\mathbf{k})$ are given in (3.1) and (3.4), respectively. The wave function in the image plane is given by the inverse Fourier transform of the aberrated wave function:

$$\Psi_{\text{image}}(\mathbf{r}) = \int\limits_{-\infty}^{\infty} \sum_n t(\mathbf{k})F(\mathbf{g}_n)\delta(\mathbf{k} - \mathbf{g}'_n)\exp(-2\pi i\mathbf{k}\cdot\mathbf{r})\,d^2\mathbf{k}$$

$$= \sum_n t(\mathbf{g}'_n)F(\mathbf{g}_n)\exp(-2\pi i\mathbf{g}'_n\cdot\mathbf{r})\,, \tag{3.11}$$

where the sum extends over all beams passing through the objective lens aperture. Using (3.11), the image intensity is given by

$$I(\mathbf{r}) = \Psi_{\text{image}}(\mathbf{r})\Psi^*_{\text{image}}(\mathbf{r})$$

$$= \sum_n \sum_m F(\mathbf{g}_n)F^*(\mathbf{g}_m)t(\mathbf{g}'_n)t^*(\mathbf{g}'_m)\exp(-2\pi i(\mathbf{g}'_n - \mathbf{g}'_m)\cdot\mathbf{r})\,.$$

$$\tag{3.12}$$

Let us now consider the diffractogram $\tilde{I}(\mathbf{k}) = \mathcal{F}I(\mathbf{r})$:

$$\tilde{I}(\mathbf{k}) = \sum_n \sum_m F(\mathbf{g}_n)F^*(\mathbf{g}_m)\,t(\mathbf{g}'_n)t^*(\mathbf{g}'_m)$$

$$\times \int\limits_{-\infty}^{\infty} \exp\left[-2\pi i(\mathbf{g}'_n - \mathbf{g}'_m - \mathbf{k})\cdot\mathbf{r}\right]d^2\mathbf{r}$$

$$= \sum_n \sum_m F(\mathbf{g}_n)F^*(\mathbf{g}_m)\,t(\mathbf{g}'_n)t^*\,(\mathbf{g}'_m)\delta(\mathbf{g}_n - \mathbf{g}_m - \mathbf{k})\,. \tag{3.13}$$

A comparison of (3.1.2) with (3.9) clearly shows the difference between the diffraction pattern and the diffractogram. Each reflection $k = \mathbf{g}_p$ in the diffractogram is caused by the interference of two beams $\mathbf{g}_n$ and $\mathbf{g}_m$ that fulfill the condition $\mathbf{g}_n - \mathbf{g}_m = \mathbf{g}_p$. As an example, the (002) reflection in the diffractogram results from the interference of (002) with (000), plus (004) with (002), plus (006) with (004), ... The amplitude $J(\mathbf{g}_p)$ of the reflection $k = \mathbf{g}_p$ is given by

$$\boxed{J(\mathbf{g}_p) = \sum_m F(\mathbf{g}_p + \mathbf{g}_m)F^*(\mathbf{g}_m)t(\mathbf{g}_p + \mathbf{g}'_m)t^*(\mathbf{g}'_m)\,.} \tag{3.14}$$

The whole diffractogram is given, according to (3.1.2), by

$$\tilde{I}(\mathbf{k}) = \sum_p J(\mathbf{g}_p)\delta(\mathbf{g}_p - \mathbf{k})\,. \tag{3.15}$$

Note that this result does not take effects of incoherence into account. The term of the sum in (3.14) with $\mathbf{g}_m = 0$ is called the *linear contribution*. In contrast, those terms with $\mathbf{g}_m \neq 0$ are called *nonlinear contributions*. For sufficiently small specimen thicknesses, the linear contributions are dominant because they stem from interference with the strong, undiffracted beam $\mathbf{g}_m = 0$.

It is instructive to see that an integral formulation of (3.14) can be readily obtained from (3.5), (3.6) and (3.10) by application of the convolution theorem (see Sect. B.2) as follows:

$$\widetilde{I}(\mathbf{k}) = \mathcal{F} \left| \mathcal{F}^{-1} \left[t \mathcal{F} \Psi \right] \right|^2 = \mathcal{F} \left[(\mathcal{F}^{-1} t \otimes \Psi)(\mathcal{F}^{-1} t^\star \otimes \Psi^\star) \right]$$
$$= \mathcal{F}(\mathcal{F}^{-1} t \otimes \Psi) \otimes \mathcal{F}(\mathcal{F}^{-1} t^\star \otimes \Psi^\star)$$
$$= (t\widetilde{\Psi}) \otimes (t^\star \widetilde{\Psi}^\star)$$
$$= \int \widetilde{\Psi}(\mathbf{k} + \mathbf{k}') \, \widetilde{\Psi}^\star(\mathbf{k}') \, t(\mathbf{k} + \mathbf{k}') \, t^\star(\mathbf{k}') \, \mathrm{d}^2 \mathbf{k}' \, , \tag{3.16}$$

where $\widetilde{\Psi}(\mathbf{k})$ is the Fourier transform of $\Psi(\mathbf{r})$ and $\otimes$ denotes a convolution.

3.2 Effects of Incoherence

In the previous section, the coherent nonlinear imaging of a crystalline specimen was described. This section describes how the effects of temporal and spatial incoherence are taken into account. Temporal incoherence arises from the (thermal) energy distribution of the electrons emitted from the cathode, fluctuations of the high tension used to accelerate the electrons of the incident beam and fluctuations of the objective lens current. Spatial incoherence arises from different propagation directions of the incident electrons caused by the finite extent of the effective electron source. In microscopes where the diameter of the filament is larger than 1 micrometer (as is the case for a hairpin filament, a pointed filament and some LaB_6 emitters), the second condenser aperture is incoherently filled. This means that each point inside the aperture can be regarded as an emitter of electrons. Since "one electron can interfere only with itself", waves starting from different positions in the condenser aperture are taken to be incoherent. Each point in the emitter thus gives rise to an individual image, and the total intensity distribution is obtained by summing the intensities of the individual images. Note that this is completely different from the interference of coherent waves, where the individual wave functions sum up instead of the intensities. The illuminating electrons impinging on the specimen form a cone with a half-angle α_C that describes the spatial incoherence. The effect of spatial incoherence is small for a microscope equipped with a field emission gun (FEG). In this case, the second condenser lens cannot be regarded as incoherently filled anymore. The good coherence of an FEG is expressed by its small convergence angle $\alpha_C \leq 5 \times 10^{-4}$ rad. In the following two sections, we shall treat the effect of spatial incoherence on image formation first and then discuss the impact of temporal incoherence.

3.2.1 Spatial Incoherence

As mentioned above, spatial incoherence is due to a distribution of incidence directions that deviate slightly from the mean direction. An electron wave that is tilted with respect to the specimen surface gives rise to a tangential component $\mathbf{k}_0^{\|}$ of its wave vector, as shown in Fig. 2.8b. Note that the normal to the specimen surface is assumed to be parallel to the symmetry axis of the illumination cone. To describe the direction of tilt, we define the "tilt angle" as a dimensionless vector $\mathbf{q}$ as follows:

$$\mathbf{q} := \lambda \mathbf{k}_0^{\|} \, . \tag{3.17}$$

We assume a Gaussian distribution of angles $\mathbf{q}$ in the illumination cone, expressed by

$$f_S(q) = \frac{1}{\pi \alpha_C^2} \exp\left(-\frac{\mathbf{q} \cdot \mathbf{q}}{\alpha_C^2}\right) \, . \tag{3.18}$$

Equation (3.18) indicates that small tilt angles are more probable than large angles. The distribution $f_S(q)$ fulfills the condition $\int_{-\infty}^{\infty} f_S(\mathbf{q}) \, \mathrm{d}^2 \mathbf{q} = 1$. Now, all individual images corresponding to different tilt angles $\mathbf{q}$ are incoherently summed:

$$I_S(\mathbf{r}) = \int\limits_{-\infty}^{\infty} I(\mathbf{r}; \mathbf{q}) f_S(q) \, \mathrm{d}^2 \mathbf{q} \, , \tag{3.19}$$

where the dependence of the image intensity on $\mathbf{q}$ has been expressed explicitly in $I(\mathbf{r}; \mathbf{q})$. Using (3.19), the diffractogram becomes

$$\tilde{I}_S(\mathbf{k}) = \mathcal{F} I_S(\mathbf{r}) = \int\limits_{-\infty}^{\infty} [\mathcal{F} I(\mathbf{r}; \mathbf{q})] f_S(q) \, \mathrm{d}^2 \mathbf{q} = \int\limits_{-\infty}^{\infty} \tilde{I}(\mathbf{k}; \mathbf{q}) f_S(q) \, \mathrm{d}^2 \mathbf{q} \, . \tag{3.20}$$

The amplitudes of the diffractogram reflections (3.14) are given by

$$J_S(\mathbf{g}_p)$$
$$= \int\limits_{-\infty}^{\infty} f_S(q) \sum_m F(\mathbf{g}_p + \mathbf{g}_m) F^*(\mathbf{g}_m) t \left(\mathbf{g}_p + \mathbf{g}_m + \frac{\mathbf{q}}{\lambda}\right) t^* \left(\mathbf{g}_m + \frac{\mathbf{q}}{\lambda}\right) \mathrm{d}^2 \mathbf{q} \, ,$$
$$\tag{3.21}$$

where $\mathbf{g'}_m = \mathbf{g}_m + \mathbf{q}/\lambda$ has been used, in accordance with (3.8) and (3.17). Note that we have assumed in (3.21) that the amplitudes of the beams do not depend on $\mathbf{q}$. This is a good approximation for the small convergence angles of less than 0.5 mrad considered here. For small convergence angles α_C we may use a first-order approximation of the wave aberrations (3.10):

$$\chi\left(\mathbf{g}_m + \frac{\mathbf{q}}{\lambda}\right) \cong \chi(\mathbf{g}_m) + \nabla \chi(\mathbf{g}_m) \cdot \frac{\mathbf{q}}{\lambda} \, . \tag{3.22}$$

Using (3.22) and (3.10), we obtain

$$t \left(\mathbf{g}_p + \mathbf{g}_m + \frac{\mathbf{q}}{\lambda} \right) t^* \left(\mathbf{g}_m + \frac{\mathbf{q}}{\lambda} \right)$$

$$= t(\mathbf{g}_p + \mathbf{g}_m) t^*(\mathbf{g}_m) \exp \left\{ i \left[\nabla \chi (\mathbf{g}_p + \mathbf{g}_m) - \nabla \chi (\mathbf{g}_m) \right] \cdot \frac{\mathbf{q}}{\lambda} \right\} . \tag{3.23}$$

Inserting (3.23) into (3.21) and using

$$E_S(\mathbf{g}_p + \mathbf{g}_m, \mathbf{g}_m; \Delta f)$$

$$:= \int_{-\infty}^{\infty} f_S(q) \exp \left\{ i \left[\nabla \chi (\mathbf{g}_p + \mathbf{g}_m) - \nabla \chi (\mathbf{g}_m) \right] \cdot \frac{\mathbf{q}}{\lambda} \right\} d^2 \mathbf{q} , \tag{3.24}$$

we finally obtain

$$J_S(\mathbf{g}_p) = \sum_m F(\mathbf{g}_p + \mathbf{g}_m) F^*(\mathbf{g}_m) t(\mathbf{g}_p + \mathbf{g}_m) t^*(\mathbf{g}_m) E_S(\mathbf{g}_p + \mathbf{g}_m, \mathbf{g}_m; \Delta f) ,$$

$$\tag{3.25}$$

where

$$\boxed{E_S(\mathbf{g}_p + \mathbf{g}_m, \mathbf{g}_m; \Delta f) = \exp \left\{ -\frac{1}{4} \frac{\alpha_C^2}{\lambda^2} \left[\nabla \chi (\mathbf{g}_p + \mathbf{g}_m) - \nabla \chi (\mathbf{g}_m) \right]^2 \right\} .}$$

$$\tag{3.26}$$

In the derivation of (3.26) we have used the relationship

$$\int_{-\infty}^{\infty} \exp \left(\frac{-x^2}{a^2} \right) \exp(ibx) \, dx = |a| \sqrt{\pi} \exp \left(-\frac{1}{4} a^2 b^2 \right) . \tag{3.27}$$

3.2.2 Temporal Incoherence

We now turn to the treatment of temporal incoherence. All three sources of temporal incoherence, i.e. the fluctuations of the lens current and of the high tension, and the energy spread of the electrons leaving the filament, can be described by a (time-dependent) variation of the defocus

$$\Delta f' = \Delta f + \epsilon \tag{3.28}$$

during the exposure of the image. A fluctuation of the current in the objective lens results in a variation of its focal length, thus changing the defocus value. Similarly, both the instability of the high tension and the electron energy spread affect the effective defocus because the focal length of the objective lens depends on the wavelength of the electrons. All these effects can be collected together in a mean *"focal spread"* parameter Δ describing the standard deviation of an assumed Gaussian distribution of defocus values Δf,

$$f_T(\epsilon) = \frac{1}{\sqrt{2\pi}\Delta} \exp\left(-\frac{\epsilon^2}{2\Delta^2}\right) . \tag{3.29}$$

Note that this definition is not unique in the literature. The EMS program package [8], for example, uses a defocus spread parameter given by $\sqrt{2}\Delta$. All images corresponding to fluctuating defocus values are summed incoherently. In analogy to (3.21), the amplitudes of the reflections in the resulting diffractogram are given by

$$J_T(\mathbf{g}_p) = \int_{-\infty}^{\infty} f_T(\epsilon) \sum_m F(\mathbf{g}_p + \mathbf{g}_m) F^*(\mathbf{g}_m)$$

$$\times\, t(\mathbf{g}_p + \mathbf{g}_m; \Delta f + \epsilon) t^*(\mathbf{g}_m; \Delta f + \epsilon)\, d\epsilon , \tag{3.30}$$

where the dependence of the coherent phase shifts $t(\mathbf{g}_m)$ on the defocus has been indicated explicitly using $t(\mathbf{g}_m; \Delta f)$. Again, we use a first-order approximation of the wave aberrations (see (3.10)):

$$\chi(\mathbf{g}_m; \Delta f + \epsilon) \cong \chi(\mathbf{g}_m; \Delta f) + \frac{\partial\chi(\mathbf{g}_m; \Delta f)}{\partial\Delta f}\, \epsilon , \tag{3.31}$$

where the dependence on the defocus has once again been indicated explicitly. The phase shifts t in (3.30) now become

$$t(\mathbf{g}_p + \mathbf{g}_m; \Delta f + \epsilon) t^*(\mathbf{g}_m; \Delta f + \epsilon) = t(\mathbf{g}_p + \mathbf{g}_m; \Delta f) t^*(\mathbf{g}_m; \Delta f)$$

$$\times \exp\left\{ i\left[\frac{\partial\chi(\mathbf{g}_p + \mathbf{g}_m; \Delta f)}{\partial\Delta f} - \frac{\partial\chi(\mathbf{g}_m; \Delta f)}{\partial\Delta f}\right]\epsilon\right\} . \tag{3.32}$$

Carrying out the integration in (3.30) using

$$E_T(\mathbf{g}_p + \mathbf{g}_m, \mathbf{g}_m; \Delta f)$$

$$:= \int_{-\infty}^{\infty} f_T(\epsilon) \exp\left\{ i\left[\frac{\partial\chi(\mathbf{g}_p + \mathbf{g}_m; \Delta f)}{\partial\Delta f} - \frac{\partial\chi(\mathbf{g}_m; \Delta f)}{\partial\Delta f}\right]\epsilon\right\} d\epsilon , \tag{3.33}$$

we finally obtain

$$J_T(\mathbf{g}_p)$$

$$= \sum_m F(\mathbf{g}_p + \mathbf{g}_m) F^*(\mathbf{g}_m) t(\mathbf{g}_p + \mathbf{g}_m) t^*(\mathbf{g}_m) E_T(\mathbf{g}_p + \mathbf{g}_m, \mathbf{g}_m; \Delta f) ,$$

$$\tag{3.34}$$

where

$$\boxed{\begin{aligned} &E_T(\mathbf{g}_p + \mathbf{g}_m, \mathbf{g}_m; \Delta f) \\ &= \exp\left\{ -\frac{1}{2}\Delta^2 \left[\frac{\partial\chi(\mathbf{g}_p + \mathbf{g}_m; \Delta f)}{\partial\Delta f} - \frac{\partial\chi(\mathbf{g}_m; \Delta f)}{\partial\Delta f}\right]^2\right\} . \end{aligned}} \tag{3.35}$$

Here, we have again made use of (3.27).

3.2.3 The Transmission Cross-Coefficients

Collecting together the results of the previous two sections, we can express the amplitude of a reflection $\mathbf{g}_p$ in the diffractogram of the image as

$$J(\mathbf{g}_p) = \sum_m F(\mathbf{g}_p + \mathbf{g}_m)F^\star(\mathbf{g}_m)T(\mathbf{g}_p + \mathbf{g}_m, \mathbf{g}_m; \Delta f) , \qquad (3.36)$$

where the $T(\mathbf{g}_p + \mathbf{g}_m, \mathbf{g}_m; \Delta f)$ are called the *transmission cross-coefficients*. They are obtained from (3.10), (3.26) and (3.35) as follows:

$$
\begin{aligned}
&T(\mathbf{g}_p + \mathbf{g}_m, \mathbf{g}_m; \Delta f) \\
&= t(\mathbf{g}_p + \mathbf{g}_m)t^\star(\mathbf{g}_m)E_S(\mathbf{g}_p + \mathbf{g}_m, \mathbf{g}_m; \Delta f)E_T(\mathbf{g}_p + \mathbf{g}_m, \mathbf{g}_m; \Delta f) .
\end{aligned}
$$
$$(3.37)$$

The transmission cross-coefficients have the properties

$$T(\mathbf{g}_p + \mathbf{g}_m, \mathbf{g}_m; \Delta f) = T^\star(\mathbf{g}_m, \mathbf{g}_p + \mathbf{g}_m; \Delta f) , \qquad (3.38)$$

$$T(-\mathbf{g}_p - \mathbf{g}_m, -\mathbf{g}_m; \Delta f) = T(\mathbf{g}_p + \mathbf{g}_m, \mathbf{g}_m; \Delta f) , \qquad (3.39)$$

$$T(\mathbf{g}_l, \mathbf{g}_l; \Delta f) = 1 . \qquad (3.40)$$

Making use of (3.4), (3.10) and (3.35), we can express the contribution of the temporal incoherence as

$$E_T(\mathbf{g}_p + \mathbf{g}_m, \mathbf{g}_m; \Delta f) = \exp\left\{ -\frac{1}{2}\pi^2 \Delta^2 \lambda^2 \left[(\mathbf{g}_p + \mathbf{g}_m)^2 - \mathbf{g}_m^2 \right]^2 \right\} . \qquad (3.41)$$

Obviously, the effects of incoherence yield a damping of the transmission cross-coefficients, and the $t(\mathbf{g}_m)$ describe a phase shift of the electron wave. Note that (3.41) reveals that it is not possible to factorize E_T in the form $E_T(\mathbf{g}_p + \mathbf{g}_m, \mathbf{g}_m; \Delta f) = E_T'(\mathbf{g}_p + \mathbf{g}_m)E_T''(\mathbf{g}_m)$, because of the cross-term in (3.41). As a consequence, image formation cannot be described by an equation analogous to (3.5) when effects of incoherence are taken into account. This has stringent consequences for the numerical computation of images, where the use of a fast Fourier transform (FFT) allows a fast image calculation if (3.5) is used. In practice, a fast computation [9] is possible in the case of a sufficiently small beam convergence angle, as occurs in instruments equipped with a an FEG. In this case, the cross-terms in (3.26) can be neglected. The temporal coherence is taken into account by computation of images for different defocus deviations ϵ and summing the resulting image intensities [9].

3.2.4 Delocalization

Delocalization is one of the major sources of error that affect the quantitative analysis of HRTEM images. Figure 3.2 explains the effect. Owing to objective

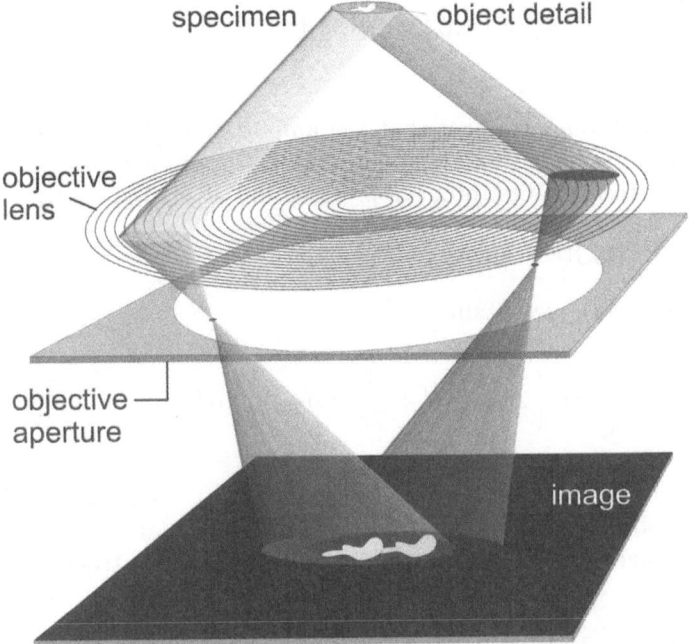

specimen   object detail

objective
lens

objective
aperture

image

Fig. 3.2. Schematic drawing explaining the effect of delocalization. At the top of
the ray path, an object detail is shown. The image detail is transferred to the image
plane by two diffracted beams. In the image plane, each beam generates an image
of the object detail. Owing to delocalization, the two images of the object detail do
not coincide, but are *delocalized* [Please see Plate 2 for color reproduction of this
plate]

lens aberrations, an object detail that is transferred to the image plane by
two beams yields images that do not coincide but instead are separated.
In this section, we calculate the amount of delocalization. For this purpose,
we consider a wave function $\Psi_n(\mathbf{r})$ that corresponds to $\mathbf{g}_n$ in Fourier space.
We assume that $\Psi_n(\mathbf{r})$ originates from a real specimen that contains edges,
interfaces or some other significant feature that sets it apart from a perfect
crystal, so that $\mathcal{F}\Psi_n(\mathbf{r})$ is not a δ-"function", but can still be described as a
function that is peaked at $\mathbf{k} = \mathbf{g}_n$ in Fourier space and has a rather small
half-width. In the diffraction pattern at the back focal plane of the objective
lens, the wave function $\widetilde{\Psi}_n(\mathbf{k})$ is the Fourier transform of $\Psi_n(\mathbf{r})$:

$$\widetilde{\Psi}_n(\mathbf{k}) = \int\limits_{-\infty}^{\infty} \Psi_n(\mathbf{r}') \exp\left(2\pi i \mathbf{k} \cdot \mathbf{r}'\right) \mathrm{d}^2\mathbf{r}' \,. \tag{3.42}$$

Owing to objective lens aberrations and defocus, a phase shift $\chi(\mathbf{k})$ is imposed
on $\widetilde{\Psi}_n(\mathbf{k})$, and we obtain

$$\widetilde{\Psi}'_n(\mathbf{k}) = \exp\{i\chi(\mathbf{k})\} \int\limits_{-\infty}^{\infty} \Psi_n(\mathbf{r}') \exp(2\pi i \mathbf{k} \cdot \mathbf{r}') \, d^2\mathbf{r}' \ . \tag{3.43}$$

We now take into account the fact that the amplitude of $\widetilde{\Psi}'_n(\mathbf{k})$ differs significantly from zero in only a small region around $\mathbf{k} = \mathbf{g}_n$ in Fourier space, so that it is sufficient to use a first-order approximation of $\chi(\mathbf{k})$ given by

$$\chi(\mathbf{k}) \approx \chi(\mathbf{g}_n) + \nabla\chi(\mathbf{g}_n)(\mathbf{k} - \mathbf{g}_n) \ . \tag{3.44}$$

Inserting (3.44) into (3.43), we obtain

$$\widetilde{\Psi}'_n(\mathbf{k}) = C \int\limits_{-\infty}^{\infty} \Psi_n(\mathbf{r}') \exp\left\{2\pi i \mathbf{k} \cdot \left[\mathbf{r}' + \frac{1}{2\pi}\nabla\chi(\mathbf{g}_n)\right]\right\} d^2\mathbf{r}' \ , \tag{3.45}$$

where the constant phase factor is given by $C = \exp\{i\left[\chi(\mathbf{g}_n) - \nabla\chi(\mathbf{g}_n) \cdot \mathbf{g}_n\right]\}$. The transformation of the wave function from the back focal plane of the objective lens to the image plane is calculated by an inverse Fourier transform:

$$\Psi'_n(\mathbf{r})$$

$$= C \int\limits_{-\infty}^{\infty} \Psi_n(\mathbf{r}') \left\{ \int\limits_{-\infty}^{\infty} \exp\left[2\pi i \mathbf{k} \cdot \left(\mathbf{r}' + \frac{1}{2\pi}\nabla\chi(\mathbf{g}_n) - \mathbf{r}\right)\right] d^2\mathbf{k} \right\} d^2\mathbf{r}'$$

$$= C \int\limits_{-\infty}^{\infty} \Psi_n(\mathbf{r}') \delta\left[\mathbf{r} - \left(\mathbf{r}' + \frac{1}{2\pi}\nabla\chi(\mathbf{g}_n)\right)\right] d^2\mathbf{r}'$$

$$= C\Psi_n\left(\mathbf{r} - \frac{1}{2\pi}\nabla\chi(\mathbf{g}_n)\right) \ . \tag{3.46}$$

Apart from the phase shift C, the wave function $\Psi'_n(\mathbf{r})$ in the image plane is shifted by an amount $1/2\pi\nabla\chi(\mathbf{g}_n)$ with respect to the unaberrated wave function $\Psi_n(\mathbf{r})$. The aberration function $\chi(\mathbf{k})$ can thus be interpreted as a phase wedge in a small region around $\mathbf{k} = \mathbf{g}_n$ that leads to a shift in real space. This is analogous to the imaging of a magnetic specimen, where the magnetic field causes a phase wedge in real space that leads a shift of the beams in the diffraction pattern. Figure 3.3 shows the amount of delocalization to be expected in a Philips CM200/ST FEG microscope with a spherical-aberration constant $C_S = 1.2$ mm. Figure 3.3a displays the aberration function $\chi(k)$ plotted versus the coordinate k in reciprocal space. Commonly used defoci Δf in the range of -200 nm to 200 nm yield a delocalization of a few nanometers, which increases strongly with increasing spatial frequency $\mathbf{k}$. Zero delocalization occurs for certain spatial frequencies at negative values of the defocus (underfocus) (Fig. 3.3b), where $\chi(k)$ has minima as shown in Fig. 3.3a. The *central beam*, which is parallel to the optic axis of the microscope and corresponds to $\mathbf{k} = 0$, is not affected by delocalization. Note that the central beam

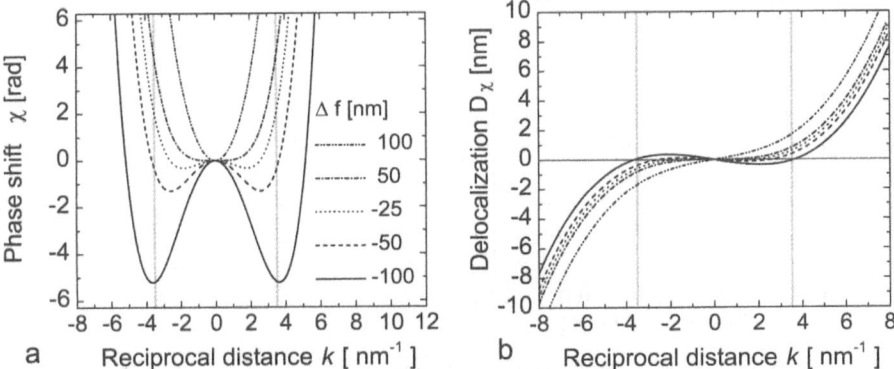

Fig. 3.3. (a) Phase shift $\chi(k)$ due to spherical aberration and defocus Δf, plotted versus the coordinate k in reciprocal space. A spherical-aberration constant of $C_S = 1.2$ mm has been assumed. The *vertical gray lines* correspond to the position of the (002) beam in GaAs. (b) Delocalization of beams as a function of Δf

is not necessarily the same as the undiffracted beam, because the electron beam can be tilted in the microscope before it reaches the specimen. The defocus value that is associated with zero delocalization of a certain beam $\mathbf{g}_n$ with respect to the central beam can be obtained from the condition $\nabla\chi(\mathbf{g}_n) = 0$, which yields

$$\Delta f_0 = -C_S\lambda^2\mathbf{g}_n^2 \ . \tag{3.47}$$

It is of special interest here to consider the spatial frequency that corresponds to the (002) lattice planes in GaAs. Minimum delocalization is obtained for $C_S = 1.2$ mm at a defocus value of $\Delta f_0 = -94.4$ nm, as displayed in Fig. 3.4. Note that only certain beams are transferred without delocalization. These beams include the beam on the optic axis and those lying on a circle in reciprocal space where $\chi(\mathbf{k})$ has its minimum. Delocalized beams interfere in the image plane. In the case of a ZA orientation of the specimen, the delocalization effect leads to images of interfaces that appear broadened in the HRTEM image. Another useful picture describing delocalization can be derived from 3.5, which describes imaging without effects of incoherence. Owing to the relation

$$I(\mathbf{r}) = \mathcal{F}^{-1}\left\{\exp\left[i\chi(\mathbf{k})\right]\mathcal{F}\Psi(\mathbf{r})\right\} = \mathcal{F}^{-1}\exp\left[i\chi(\mathbf{k})\right] \otimes \Psi(\mathbf{r}) \ , \tag{3.48}$$

where $\otimes$ represents the convolution operation, each point of the wave function $\Psi(\mathbf{r})$ is spread into an image described by its convolution with the *point spread function* $\mathcal{F}^{-1}\exp\left[i\chi(\mathbf{k})\right]$. This description is especially useful in connection with channeling theory, whereas the picture of delocalized reflections is useful when single reflections are considered.

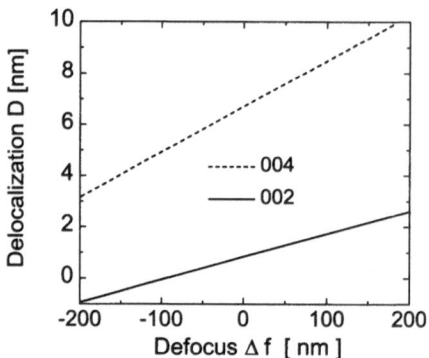

Fig. 3.4. Delocalization of the (002) and (004) beams in GaAs plotted versus the defocus Δf. A spherical-aberration constant $C_S = 1.2$ mm has been used

3.3 Reconstruction of the Electron Wave Function

It was pointed out earlier that the interpretation and quantitative evaluation of HRTEM images is hampered by the loss of the phase of the electron wave in the image plane, where only the intensity can be recorded. In this section, two methods are briefly outlined which allow the reconstruction of the electron wave function at the object exit surface.

The first procedure is based on the acquisition of a defocus series instead of only one HRTEM image. This method is called the *focal-variation reconstruction method* [9]. The method requires a knowledge of the wave aberrations that are present during the exposure of the defocus series. The focal-variation reconstruction method appears to be especially useful when the intended image interpretation is based on channeling theory, i.e. when a high-resolution image pattern is used to obtain the positions of atomic columns in the specimen. The reconstructed wave function is already corrected with regard to wave aberrations induced by the objective lens, so that effects such as the delocalization are strongly diminished. Thus, intensity maxima in the reconstructed wave function allow the determination of the positions of atomic columns. Another method for wave function reconstruction, not discussed here, is the *parabola method* [10].

The second approach retrieves the phase of the electron wave by recording an *off-axis hologram* in the image plane. The hologram is formed by the interference of an electron wave that has passed through the specimen (the *object wave*) with a spatially homogeneous *reference wave*. The image intensity shows a fringe pattern, where the positions of the fringes depend on the phase difference between the object wave and the reference wave. This method does not require a knowledge of the objective lens aberrations if only a reconstruction of the electron wave in the image plane is desired. However, with a knowledge of the phase of the electron wave in the image plane, a reconstruction of the wave function at the object exit surface is straightforward if the aberrations are known. A drawback of this approach is the small signal-to-noise ratio of the reconstructed wave function. One can decrease the noise level due to electron statistics by increasing the image intensity, or one may raise the level of the "true" signal by increasing the spatial coherence that determines the contrast of the hologram fringes. Unfortunately, these approaches have competing requirements. Thus, the exposure time has to be increased, which requires a high stability of the TEM, and this is generally not satisfactory at present. Satisfactory stability demands not only a high stability of the electronic components but also the suppression of stray electromagnetic fields, a constant room temperature, low air currents and low noise, as well as improved specimen manipulators, where piezoelectric components, for example, could be applied. There is much potential to improve the stability of electron microscopes in the near future. A clear advantage of electron holography is that the reconstructed wave function includes only elastic contributions because the smallest energy difference between the reference wave and the object wave destroys their coherence. Another advantage compared with the focal-variation method is the possibility of phase detection at low spatial frequencies. Electron holography therefore also allows the measurement of mean inner potentials, electric fields, magnetic fields of small particles, etc. The possibility to measure the projected mean inner potential is particularly valuable for composition analysis using a chemically sensitive reflection because it allows an accurate determination of the local specimen thickness in the electron beam direction.

3.3.1 Focal-Variation Reconstruction

The maximum-likelihood (MAL) method for focal-variation image reconstruction in HRTEM presented in [9] seems to be the state of the art at present. In this section, a rather rough idea of its principle is given. The method is based on the acquisition of a defocus series of, for example, $N = 20$ images with defoci

$$\Delta f^{(j)} = \Delta f_0 + j\epsilon \,, \tag{3.49}$$

where ϵ is the *defocus step size*, with a value of a few nanometers, and Δf_0 is the *starting defocus*. The basic idea is to search for a wave function $\Psi(\mathbf{r})$ that show the best agreement with the experimental images. For that purpose, a simulated defocus series is calculated using $\Psi(\mathbf{r})$ and the parameters of the microscope and of the imaging process corresponding to the experimental conditions. The actual comparison is carried out in Fourier space; the procedure requires minimization of the error functional

$$S^2 = \frac{1}{N} \sum_{j=1}^{N} \int \left| \delta \widetilde{I}^{(j)}(\mathbf{k}'') \right|^2 \mathrm{d}^2 \mathbf{k}'' \,, \tag{3.50}$$

where

$$\delta \widetilde{I}^{(j)}(\mathbf{k}'') = \widetilde{I}^{(j)}_{\mathrm{E}}(\mathbf{k}'') - \widetilde{I}^{(j)}(\mathbf{k}'') \tag{3.51}$$

is the difference between the diffractogram $\widetilde{I}^{(j)}_{\mathrm{E}}(\mathbf{k}'')$ of the jth image of the defocus series and the diffractogram $\widetilde{I}^{(j)}(\mathbf{k}'')$ of the simulated image. The simulated image is computed from the wave function $\widetilde{\Psi}(\mathbf{k}'') = \mathcal{F}\Psi(\mathbf{r})$ with microscope parameters prevailing during the exposure of the jth experimental image, according to (3.16), giving

$$\widetilde{I}^{(j)}(\mathbf{k}'') = \int \widetilde{\Psi}(\mathbf{k}'' + \mathbf{k}')\,\widetilde{\Psi}^*(\mathbf{k}')\,\tau^{(j)}(\mathbf{k}'' + \mathbf{k}')\,\tau^{*(j)}(\mathbf{k}')\,\mathrm{d}^2\mathbf{k}'\ , \tag{3.52}$$

where the aberration functions $\tau^{(j)}(\mathbf{k})$ are given by

$$\tau^{(j)}(\mathbf{k}) = t^{(j)}(\mathbf{k})E^{(j)}_{\mathrm{S}}(\mathbf{k}, 0; \Delta f^{(j)})\ . \tag{3.53}$$

The function $t^{(j)}(\mathbf{k})$ is given by (3.10) and the beam convergence damping envelope $E_{\mathrm{S}}(\mathbf{k}, 0; \Delta f^{(j)})$ is given by (3.26). Note that the definitions in (3.52) and (3.53) lead to the approximation $[\nabla\chi(\mathbf{k} + \mathbf{k}') - \nabla\chi(\mathbf{k}')]^2 \approx \nabla\chi^2(\mathbf{k} + \mathbf{k}') + \nabla\chi^2(\mathbf{k}')$ in the damping envelope of the transmission cross-coefficient, which is valid for the small beam convergence angles obtained in FEG transmission electron microscopes. The effect of temporal incoherence is taken into account by an explicit (weighted) sum of images of a simulated defocus series corresponding to the focal spread for each individual experimental image. The resulting diffractogram and its difference from the experimental diffractogram are given by

$$\widetilde{I}^{(j)}_{\Sigma}(\mathbf{k}) = \frac{1}{2M + 1} \sum_{m=j-M}^{j+M} f_{\mathrm{T}}\big[(m - j)\epsilon\big]\,\widetilde{I}^{(m)}(\mathbf{k}) \quad \text{and}$$

$$\delta\widetilde{I}^{(j)}_{\Sigma}(\mathbf{k}) = \widetilde{I}^{(j)}_{\mathrm{E}}(\mathbf{k}) - \widetilde{I}^{(j)}_{\Sigma}(\mathbf{k})\ , \tag{3.54}$$

where f_{T} is the Gaussian distribution of defocus values given in (3.29). For notational convenience, the focal integration (3.54) is neglected.

Minimization of the error functional (3.50) leads to the condition

$$\Gamma(\mathbf{k}) := \frac{\partial S^2}{\partial \widetilde{\Psi}(\mathbf{k})} = 0 \quad \forall\mathbf{k}\ . \tag{3.55}$$

Inserting (3.50) and (3.51) into (3.55), we obtain

$$\Gamma(\mathbf{k}) = -\frac{2}{N} \sum_{j=1}^{N} \int \delta\widetilde{I}^{(j)}(\mathbf{k}'')\widetilde{\Psi}^*(\mathbf{k} - \mathbf{k}'')\tau^{(j)}(\mathbf{k})\,\tau^{*(j)}(\mathbf{k} - \mathbf{k}'')\,\mathrm{d}^2\mathbf{k}''\ . \tag{3.56}$$

To be able to express (3.56) as a convolution, we substitute $\mathbf{k}'' = \mathbf{k} - \mathbf{k}'$, and find:

$$\Gamma(\mathbf{k}) = -\tau^{(j)}(\mathbf{k}) \frac{2}{N} \sum_{j=1}^{N} \int \widetilde{\Psi}^*(\mathbf{k}') \tau^{*(j)}(\mathbf{k}') \, \delta \widetilde{I}^{(j)}(\mathbf{k} - \mathbf{k}') \, \mathrm{d}^2\mathbf{k}' \,. \tag{3.57}$$

Finally, (3.57) can be written using the convolution operation $\otimes$:

$$\Gamma = -\tau^{(j)} \frac{2}{N} \sum_{j=1}^{N} \left[\widetilde{\Psi}^* \tau^{*(j)} \otimes \delta \widetilde{I}^{(j)} \right]$$

$$= -\tau^{(j)} \frac{2}{N} \sum_{j=1}^{N} \mathcal{F} \left[\mathcal{F}^{-1}(\widetilde{\Psi}^* \tau^{*(j)}) \, \mathcal{F}^{-1} \delta \widetilde{I}^{(j)} \right] \,, \tag{3.58}$$

where the second line makes use of a Fourier transform, which enables the application of a fast Fourier transform (FFT) to speed up the computation. Taking account of the temporal incoherence again (see (3.54)), (3.58) becomes

$$\Gamma_\Sigma = -\frac{2}{N(2M+1)}$$

$$\times \sum_{j=1}^{N} \sum_{m=j-M}^{j+M} f_\mathrm{T}\left[(m-j)\epsilon\right] \tau^{(j)} \, \mathcal{F} \left[\mathcal{F}^{-1}(\widetilde{\Psi}^* \tau^{*(m)}) \, \mathcal{F}^{-1} \delta \widetilde{I}_\Sigma^{(j)} \right] \,. \tag{3.59}$$

The main part of the procedure is an iterative solution of the problem $\Gamma(\mathbf{k}) = 0$. Starting with $\Psi_0(\mathbf{r}) = \mathrm{const.}$, the iteration from step n to $n+1$ is given by

$$\widetilde{\Psi}_{n+1}(\mathbf{k}) = \widetilde{\Psi}_n(\mathbf{k}) + \delta \widetilde{\Psi}_n(\mathbf{k}) = \widetilde{\Psi}_n(\mathbf{k}) - \frac{1}{2}\gamma_n s_n(\Gamma_{\Sigma n}(\mathbf{k})) \,, \tag{3.60}$$

where the simplest approach is to use $s_n(\Gamma_{\Sigma n}(\mathbf{k})) = \Gamma_{\Sigma n}(\mathbf{k})$. A more sophisticated method with improved convergence properties is the application of the conjugate gradient approach described in [9]. The real number γ_n is obtained by computing the error functional S^2 for three different values, e.g. $\gamma_n = 0, 1/2$ and 1, and finding the minimum of S^2 by fitting a parabola.

A drawback of focal-variation image reconstruction in practice is specimen drift during the exposure of the image series. This problem requires the alignment of the images during the reconstruction using a cross-correlation method. As a consequence, images can only be reconstructed if they contain significant features that can be safely recognized by the cross-correlation procedure. One example where this requirement is fulfilled is the evaluation of nanometer-scale semiconductor islands described in Sect. 8.2.2.3.

3.3.2 Off-Axis Electron Holography

Figure 3.5a shows the principle of off-axis holography using a Möllenstedt biprism [11]. The biprism is a thin (≈ 100 nm) metal-coated wire connected to a power supply which gives it a positive potential with respect to ground.

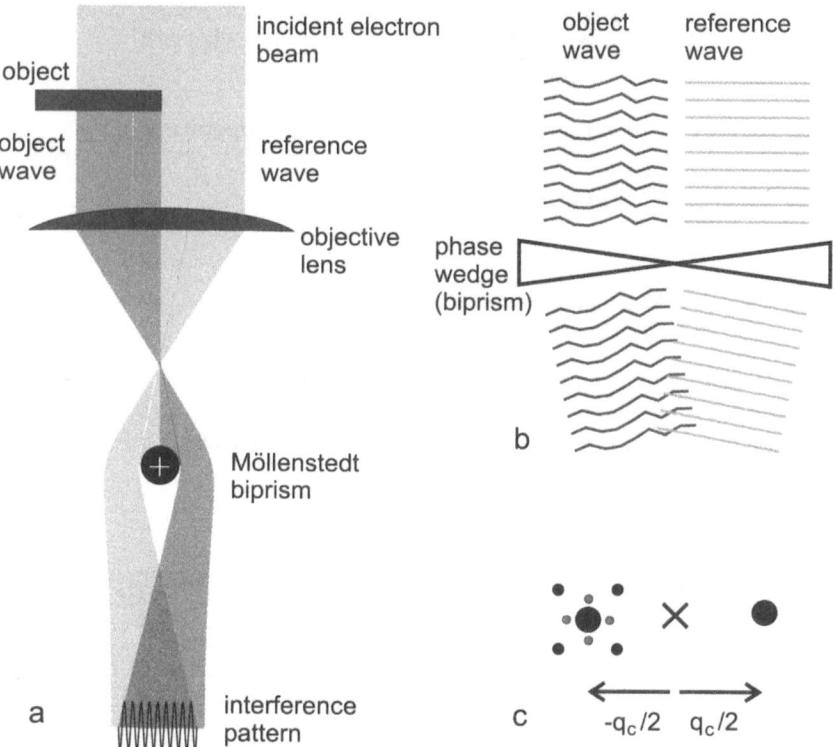

Fig. 3.5. (a) Schematic drawing explaining the principle of off-axis holography using a Möllenstedt biprism. The biprism leads to interference of the object wave and the reference wave in the image plane. (b) Because of the biprism, the object wave and the reference wave are tilted in opposite directions. The action of the biprism can thus be described by a phase wedge in real space that leads to a shift of the beams in reciprocal space. In Fourier space (c), the reflections in the object wave and in the reference wave have a distance $\mathbf{q}_c$

It is installed slightly above the first intermediate image plane, in the holder that also contains the selected-area diffraction diaphragm. The object splits the incident electron beam into a wave passing through the specimen (the object wave) and an electron wave in the vacuum adjacent to the specimen (the reference wave). The electric field induced by the biprism leads to a tilt of the object and reference waves in such a way that they overlap in the image plane. If the spatial coherence is sufficient, their interference gives a fringe pattern. In a wave-optical picture, the action of the biprism can be described by a phase wedge, as displayed in Fig. 3.5b. Assuming an object wave $\Psi_{\mathrm{Obj}}^0(\mathbf{r})$ and a reference wave $\Psi_{\mathrm{Ref}}^0(\mathbf{r}) = 1$, we obtain (see Fig. 3.5c)

$$\Psi_{\mathrm{Obj}}(\mathbf{r}) = \Psi_{\mathrm{Obj}}^0 \exp\left(2\pi i \frac{\mathbf{q_c}}{2}\cdot\mathbf{r}\right) ,$$

$$\Psi_{\mathrm{Ref}}(\mathbf{r}) = \exp\left(-2\pi i \frac{\mathbf{q_c}}{2}\cdot\mathbf{r}\right) , \tag{3.61}$$

where $\mathbf{q_c}$ describes the phase wedge and is defined by the biprism voltage. The total wave function is $\Psi(\mathbf{r}) = \Psi_{\mathrm{Obj}}(\mathbf{r}) + \Psi_{\mathrm{Ref}}(\mathbf{r})$. We now calculate a "virtual" diffraction pattern by applying a Fourier transform:

$$\widetilde{\Psi}_{\mathrm{Obj}}(\mathbf{k'}) = \int \Psi_{\mathrm{Obj}}^0(\mathbf{r}) \exp\left\{2\pi i \left(\mathbf{k'} + \frac{\mathbf{q_c}}{2}\right)\cdot\mathbf{r}\right\} \mathrm{d}^2\mathbf{r}$$

$$= \widetilde{\Psi}_{\mathrm{Obj}}^0\left(\mathbf{k'} + \frac{\mathbf{q_c}}{2}\right) ,$$

$$\widetilde{\Psi}_{\mathrm{Ref}}(\mathbf{k'}) = \int \exp\left\{2\pi i \left(\mathbf{k'} - \frac{\mathbf{q_c}}{2}\right)\cdot\mathbf{r}\right\} \mathrm{d}^2\mathbf{r}$$

$$= \delta\left(\mathbf{k'} - \frac{\mathbf{q_c}}{2}\right) . \tag{3.62}$$

As shown in Fig. 3.5, the virtual diffractogram contains the Fourier transform of the tilted object wave as well as the reflection corresponding to the reference wave, separated by $\mathbf{q_c}$. The next step is the description of the diffractogram of the hologram. For notational convenience, the origin of reciprocal space is shifted according to $\mathbf{k} = \mathbf{k'} + \mathbf{q_c}/2$. We obtain

$$\widetilde{I}(k) = \mathcal{F}\left|\mathcal{F}^{-1}\left[\widetilde{\Psi}_{\mathrm{Obj}}^0(\mathbf{k})\, t(\mathbf{k}) + \delta(\mathbf{k} - \mathbf{q_c})\right]\right|^2 , \tag{3.63}$$

where effects of incoherence have been neglected. Application of the convolution theorem yields:

$$\widetilde{I}(k) = \mathcal{F}\big[\, (\Psi_{\mathrm{Obj}}^0 \otimes (\mathcal{F}^{-1}t) + \exp(-2\pi i \mathbf{q_c}\cdot\mathbf{r})) \times$$
$$\times\, (\Psi_{\mathrm{Obj}}^{0\star} \otimes (\mathcal{F}^{-1}t^\star) + \exp(2\pi i \mathbf{q_c}\cdot\mathbf{r}))\,\big]$$

$$= \delta(\mathbf{k}) + \left(\widetilde{\Psi}_{\mathrm{Obj}}^0\, t\right) \otimes \left(\widetilde{\Psi}_{\mathrm{Obj}}^{0\star}\, t^\star\right) \qquad \text{autocorrelation}$$

$$+\, \delta(\mathbf{k} + \mathbf{q_c}) \otimes \left(\widetilde{\Psi}_{\mathrm{Obj}}^0\, t\right) \qquad \text{sideband 1}$$

$$+\, \delta(\mathbf{k} - \mathbf{q_c}) \otimes \left(\widetilde{\Psi}_{\mathrm{Obj}}^{0\star}\, t^\star\right) \qquad \text{sideband 2 .}$$

$$\tag{3.64}$$

Equation (3.64) shows that the diffractogram can be decomposed into three parts. The autocorrelation in the center is equivalent to the diffractogram of a conventional HRTEM image. Additionally, two side bands occur that are shifted by $\pm\mathbf{q_c}$ with respect to the autocorrelation. Both sidebands contain linear information about the wave function, the first one being proportional to $\widetilde{\Psi}_{\mathrm{Obj}}^0(\mathbf{k})$, and the second one being proportional to the conjugate wave function $\widetilde{\Psi}_{\mathrm{Obj}}^{0\star}(\mathbf{k})$.

Presupposing that the distance $\mathbf{q}_c$ between the bands is large enough, the sideband 1 can be selected out by filtering. To do this, a circular area is centered around the sideband 1 and all the information in the diffractogram outside the circle is deleted. In high-resolution electron holography, one generally applies the criterion

$$|\mathbf{q}_c| \geq \mathbf{k}_{\max} \tag{3.65}$$

for the separability of the sidebands, where $\mathbf{k}_{\max}$ is the maximum spatial frequency transferred by the objective lens. However, this criterion may be too strong in some special cases, particularly for the off-ZA specimen orientations chosen for the compositional analysis that is discussed later on. Once the sideband 1 has been separated, it has to be centered by a shift of $\mathbf{q}_c$. A method for centering using an empty hologram [12] that is very useful in practice is described in Sect. 9.2.2, where the application of holography to the compositional analysis of an AlAs/GaAs superlattice is described. The resulting complex image is inverse Fourier-transformed and the aberrated wave function $\Psi^0_{\mathrm{Obj}}(\mathbf{r}) \otimes \mathcal{F}^{-1} t(\mathbf{k})$ is obtained.

The treatment of effects of incoherence can be performed in analogy with Sect. 3.2. Note that the sidebands contain only contributions from elastically scattered electrons because energy losses of the order of 10^{-15} eV due to inelastic scattering processes in the specimen are enough to destroy the coherence between the object wave and the reference wave [13]. However, effects of temporal and spatial incoherence still occur in the imaging process, and we find the following for the averaged wave function in the centered diffractogram:

$$\widetilde{\Psi}^{\mathrm{aberr}}_{\mathrm{Obj}}(\mathbf{k}) = \widetilde{\Psi}^0_{\mathrm{Obj}}(\mathbf{k})\, t(\mathbf{k})\, E_{\mathrm{T}}(\mathbf{k}, 0; \Delta f)\, E_{\mathrm{S}}(\mathbf{k}, 0; \Delta f)\,. \tag{3.66}$$

The reconstructed electron wave can be corrected with respect to the coherent aberrations by a simple multiplication by $t^\star(\mathbf{k}) = \exp\{-i\chi_{\mathrm{num}}(\mathbf{k})\}$, where $\chi_{\mathrm{num}}(\mathbf{k})$ is a "numerical phase plate", as follows:

$$\widetilde{\Psi}^{\mathrm{incoh}}_{\mathrm{Obj}}(\mathbf{k}) = \widetilde{\Psi}^{\mathrm{aberr}}_{\mathrm{Obj}}(\mathbf{k}) t^\star(\mathbf{k})\,. \tag{3.67}$$

The superscript "incoh" reflects the remaining effects of incoherence. The correction of effects of incoherence is normally omitted because it would increase the noise level of the reconstructed image. The necessary accuracy of the phase plate $\chi_{\mathrm{num}}(\mathbf{k})$ for the reconstruction is given by the Rayleigh criterion [5, 14]:

$$|\chi_{\mathrm{res}}(\mathbf{k})| = |\chi_{\mathrm{mic}}(\mathbf{k}) - \chi_{\mathrm{num}}(\mathbf{k})| \leq \frac{\pi}{6} \quad \forall \mathbf{k}: \ 0 \leq |\mathbf{k}| \leq |\mathbf{k}_{\max}|\,, \tag{3.68}$$

where $\chi_{\mathrm{mic}}(\mathbf{k})$ is the actual wave aberration of the microscope, $\chi_{\mathrm{res}}(\mathbf{k})$ is the residual aberration after correction and $\mathbf{k}_{\max}$ is the maximum spatial frequency transferred.

Another effect that has not yet been addressed is the decreased amplitude of the wave function in the sidebands, which leads to a reduced signal-to-noise ratio in the reconstructed wave function. Four main reasons for this amplitude quenching can be given. First, the electrons with a small energy loss due to inelastic interaction inside the specimen do not interfere with the reference beam and give a background intensity, decreasing the contrast both in the autocorrelation and in the sidebands. Second, incomplete coherence due to beam convergence also damps the amplitude of the hologram fringes. Third, mechanical instabilities, drift of the biprism wire and stray electromagnetic fields lead to an averaging of statistically displaced holograms during the exposure time. The fourth point is the nonideal modulation transfer function of the CCD camera, which leads to a damping that increases with increasing spatial frequency in the image.

References

1. O. Scherzer: J. Appl. Phys **20**, 20 (1949)
2. J. Frank: Optik **38**, 519 (1973)
3. R.H. Wade, J. Frank: Optik **49**, 81 (1977)
4. K. Ishizuka: Ultramicroscopy **5**, 55 (1980)
5. M. Born, E. Wolf: *Principles of Optics*, 6th (corr.) edn. (Cambridge University Press, Cambridge 1998)
6. H. Alexander: *Physikalische Grundlagen der Elektronenmikroskopie*, 1st edn. (Teubner, Stuttgart 1997)
7. M. Haider, H. Rose, S. Uhlemann, E. Schwan, B. Kabius, K. Urban: Ultramicroscopy **75**, 53 (1998)
8. P.A. Stadelmann: Ultramicroscopy **51**, 131 (1987)
9. W.M.J. Coene, A. Thust, M. Op de Beeck, D. Van Dyck: Ultramicroscopy **64**, 109 (1996)
10. M. Op de Beeck, D. Van Dyck, W. Coene: Ultramicroscopy **64**, 167 (1996)
11. G. Möllenstedt, H. Düker: Z. Phys. **145**, 377 (1965)
12. M. Lehmann: *Numerische Rekonstruktion der aberrationsfreien Objektwelle aus off-axis Elektronenhologrammen*. Ph.D. Thesis, Eberhard-Karls-Universität, Tübingen (1997)
13. A. Harscher, F. Lenz, H. Lichte: "Electron holography provides zero-loss images". In: *Last Minute Brochure*, 10th European Congress on Electron Microscopy, EUREM92, Granada, Spain (1992).
14. H. Lichte: Ultramicroscopy **47**, 223 (1992)

Part II

Digital Image Analysis

Part II

Digital Image Analysis

4 Strain State Analysis

4.1 Displacements and Lattice Spacings

In this section, we shall describe the measurement of local and averaged lattice parameters and displacements. The method outlined here is similar to those suggested by Bierwolf et al. [1], Brandt et al. [2], Paciornik et al. [3], Seitz et al. [4] and Jouneau et al. [5]. It contains the following analysis steps:

1. noise reduction,
2. detection of lattice sites and gridding,
3. calculation of lattice base vectors,
4. analysis of displacements and lattice spacings.

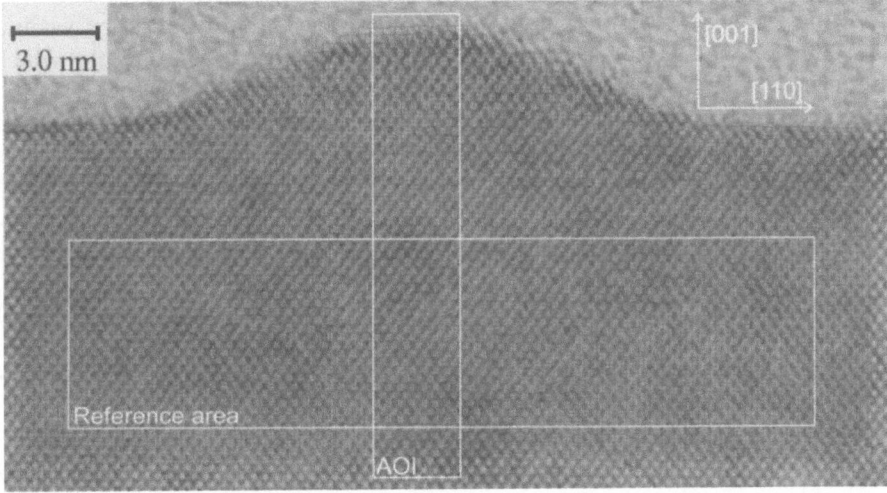

Fig. 4.1. HRTEM image of an $In_xGa_{1-x}As/GaAs(001)$ SK island containing a grid that connects the local brightness maxima of the dumbbells. The area of interest (AOI) (*blue frame*) was used for the determination of the In concentration inside the island. The reference area (*green frame*) was used for the calculation of the basis vectors of the reference lattice [Please see Plate 3 for color reproduction of this plate]

We shall describe the application of the individual analysis steps to the example HRTEM micrograph depicted in Fig. 4.1, which shows an $In_xGa_{1-x}As$ island on a GaAs(001) substrate. This cross-sectional image was taken along the [110] projection. The nominal In concentration was 60%, the nominal $In_xGa_{1-x}As$ layer thickness was 1.5 nm and the growth temperature during the MBE process was 500°C. This micrograph will also be used as an example in the description of the measurement of the local sample thickness and in the finite-element (FE) modelling. As a final result, we shall obtain a model of the local In concentration inside the island. This example was chosen because the determination of local In concentrations in the "bulk" of the island is difficult to perform by other methods with atomic-scale spatial resolution. A particular complication results from the fact that the lattice parameters parallel and perpendicular to the interface plane vary within the island owing to the elastic relaxation of the island at its free surfaces. This circumstance excludes methods where a lattice parameter fluctuation may affect the In concentration measurement; this is also the case for the lattice fringe analysis outlined in Chap. 5. We start with a description of the noise reduction procedure.

4.1.1 Noise Reduction

Images digitized with either an off-line or an on-line CCD camera that is attached directly to the microscope contain a certain amount of noise. In addition to the Poisson-distributed noise due to the electron statistics, we find thin amorphous layers on the top and bottom surfaces of the sample, which are formed during the ion milling-step of sample preparation for the TEM [6]. Other possible sources of noise are the grain of any photographic emulsion used and electronic noise in the device used for the digitizing process. For noise reduction, we have used a Wiener filtering technique [7], where the noise level is estimated locally in the Fourier-transformed image $\tilde{C}$. The Fourier-transformed image consists of the undisturbed signal $\tilde{S}$ and a noise part $\tilde{N}$:

$$\tilde{C} = \tilde{S} + \tilde{N} . \tag{4.1}$$

The noise reduction is carried out by applying a filter Φ to the Fourier-transformed image $\tilde{C}$:

$$\tilde{C}_{Nr} = \tilde{C} \times \Phi . \tag{4.2}$$

$\tilde{C}_{Nr}$ is the Fourier transform of the noise-reduced image. The filter Φ is given by

$$\Phi = \begin{cases} \dfrac{|\tilde{C}|^2 - |\tilde{N}|^2}{|\tilde{C}|^2} & \text{if } |\tilde{C}|^2 > |\tilde{N}|^2 \\ 0 & \text{otherwise} \end{cases} . \tag{4.3}$$

The filter calculated as indicated in (4.2) is often called the "optimum" or "conventional" Wiener filter. Other choices (e.g. the parametric Wiener filter) are compared in [8].

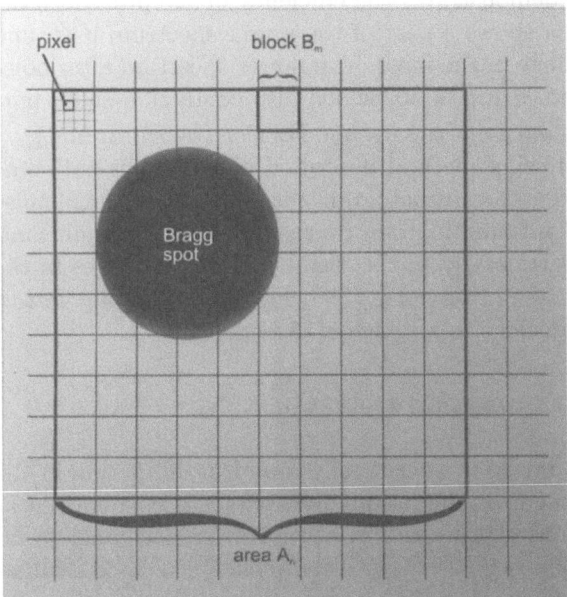

Fig. 4.2. Schematic drawing showing the decomposition of the Fourier-transformed image into areas and blocks. The smallest detectable unit is given by a pixel

In Fourier-transformed HRTEM images of defect-free lattice structures, the information is predominantly localized around those spatial frequencies which correspond to lattice spacings in real space, whereas the noise part $|\widetilde{N}|^2$ has a low, slowly varying intensity. Therefore, the noise part $|\widetilde{N}|^2$ can be estimated using the following procedure. The power spectrum $|\widetilde{C}|^2$ is divided into equally sized areas A_n (Fig. 4.2). The size of the area must exceed the sizes of the Bragg spots contained in the Fourier-transformed image. Each area is subdivided into blocks B_m. For each block, an intensity I_{B_m} is calculated as the maximum (or the mean) of all pixel intensities in B_m. For each area A_n, the intensities I_{B_m} of the blocks in A_n are averaged:

$$I_{A_n} = \frac{1}{W} \sum_{B_m} w_{B_m} I_{B_m} \; ; \quad w_{B_m} = \max_{B_m}(I_{B_m}) - I_{B_m} + 1 \; ;$$

$$W = \sum_{B_m} w_{B_m} \; . \tag{4.4}$$

The weighting factors w_{B_m} ensure that the intense Bragg peaks contribute much less ($w_{B_m} \approx 1$) to the estimation of the noise part than do blocks with low block intensities ($w_{B_m} \approx \max_{B_m}(I_{B_m})$) and that the smooth variation of the noise part inside the area is taken into account. As a result, the values I_{A_n} form a map of the noise part $|\widetilde{N}|^2$. The noise part for each pixel is calculated by bilinear interpolation with respect to I_{A_n}.

The example in Fig. 4.3 demonstrates the efficiency of the procedure described above [9]. Figure 4.3a shows a part of the power spectrum $|\widetilde{C}|^2$, and a small part of the lattice image in the inset. The power spectrum after noise reduction, $|\widetilde{C}_{Nr}|^2$, is depicted in Fig. 4.3b, which also contains a small part of the lattice image that results from the inverse Fourier transform of $\widetilde{C}_{Nr}$.

Note that the Wiener filtering method described above filters out some "true" information contained in the diffractogram that is below the local noise level. Although this kind of information can be regarded as not significant, some artifacts may occur, particularly in the vicinity of abrupt edges in the image. One artifact that is often observed is the occurrence of a faint lattice pattern in the vacuum region close to an edge of the specimen.

4.1.2 Detection of Lattice Sites and Gridding

The contrast of an HRTEM image is a result of dynamical diffraction in the crystalline specimen and depends on the sample thickness, the microscope parameters (defocus, electron energy spread and beam convergence angle) and the nonlinear image formation process. As a consequence, exact determination of the atom positions usually requires a comparison of the experimental image with simulated images. We use the positions of the intensity maxima to obtain a lattice which represents the dimensions of the projected unit cells. The positions of the intensity maxima may correspond to the locations of the columns of atoms, of the tunnel sites or – sometimes – of neither. However, our approach does not rely on a knowledge of the positions of atomic columns with respect to the intensity maxima. It is based only on the assumption of a constant spatial relationship between the positions of the intensity maxima and the columns of atoms. This requirement is often fulfilled in small areas where the specimen thickness varies insignificantly. The formation of the two-dimensional grid is performed in five steps:

1. Finding the positions $M^{(1)}$ of pixels corresponding to local brightness maxima (Fig. 4.4a).
2. Fitting parabolas to the intensity profiles along two lines L_1 and L_2 across $M^{(1)}$ that run along the x and y directions, yielding a more accurate position $M^{(2)}$ (Fig. 4.4b).
3. Fitting parabolas to the intensity profiles along four lines L_1 to L_4 (Fig. 4.4c) across $M^{(2)}$. Averaging of the positions of the maxima of the parabolas results in the final position $M^{(3)}$.
4. Formation of grid lines by connecting positions along each of two selected directions (Figs. 4.4d,e).

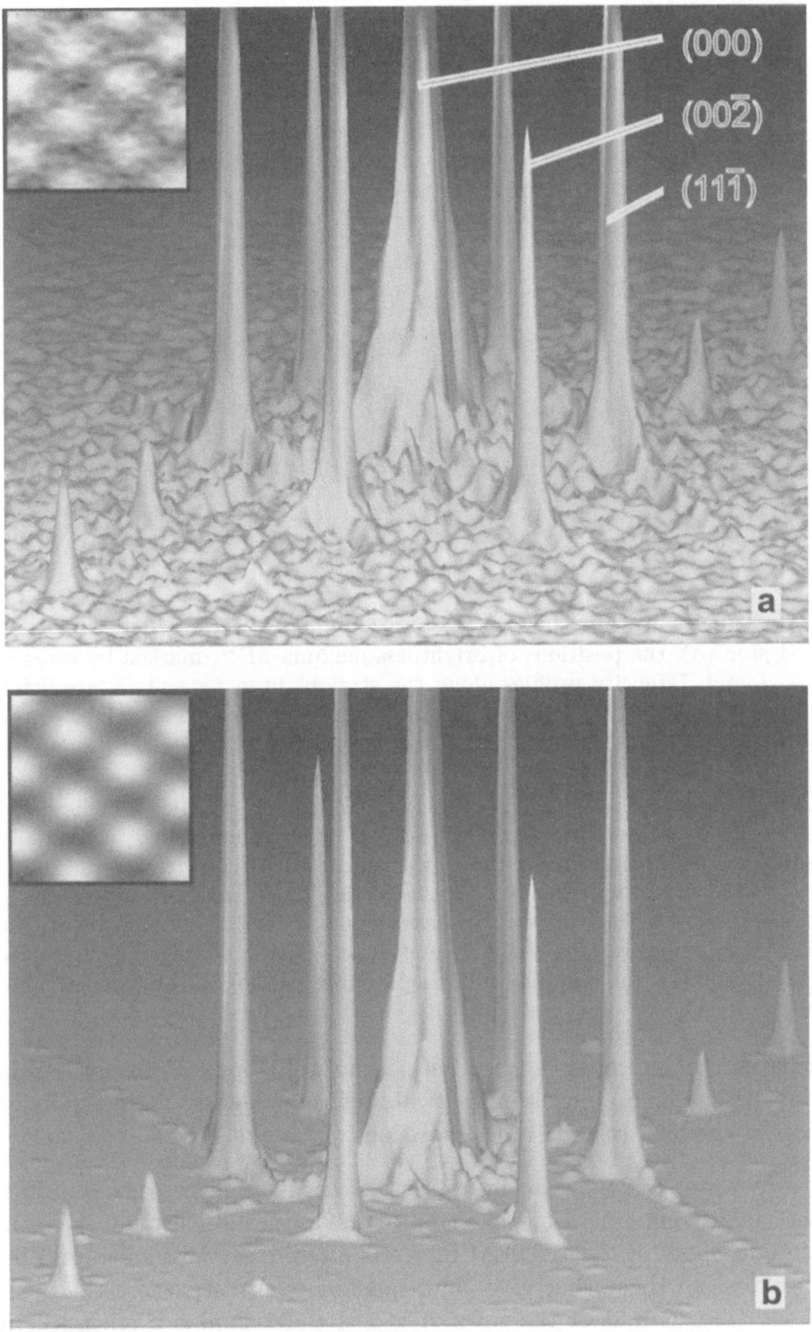

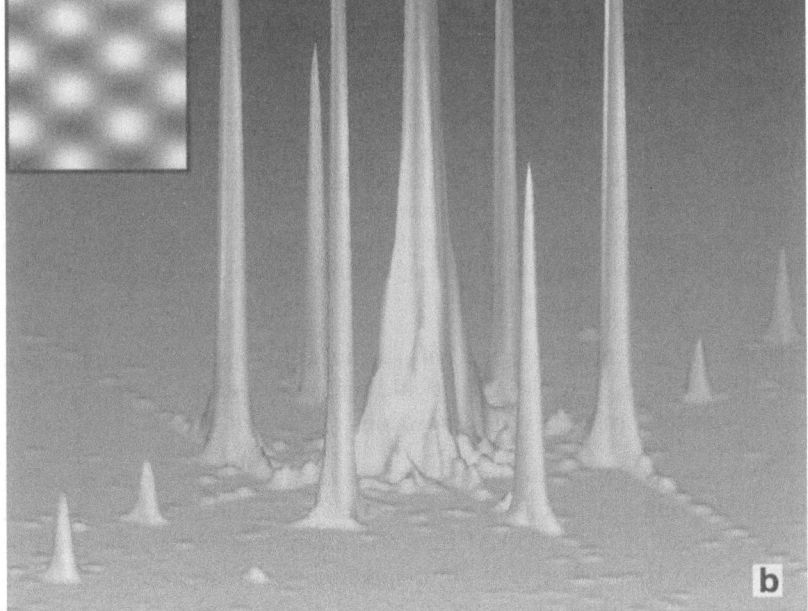

Fig. 4.3. Three-dimensional plots of the intensities of the Fourier-transformed images of (**a**) an original image and (**b**) the image after Wiener noise reduction. The *insets* show the corresponding lattice images

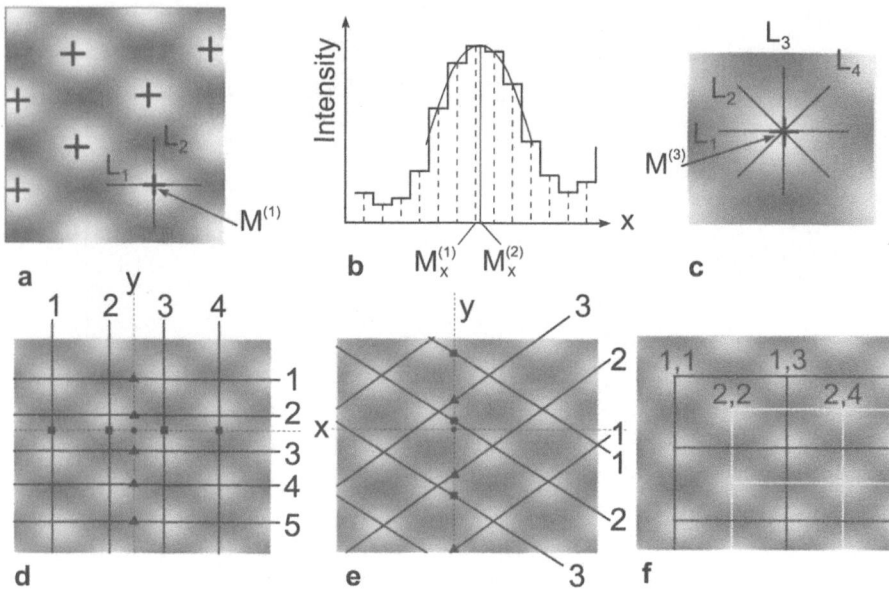

Fig. 4.4. Schematic drawing showing the procedure for lattice site determination. In the first step (**a**), the positions of brightness maxima $M^{(1)}$, marked by *crosses*, have to be found. Intensity profiles along the straight lines L_1 and L_2 are fitted by parabolas. One of these is shown in (**b**) as an example. The maxima of the parabolas, $(M_x^{(2)}, M_y^{(2)})$, give the positions of the intensity maxima with improved accuracy. The same procedure is applied to intensity profiles along $L_{1,2,3,4}$ (**c**) for further enhancement of the accuracy, yielding the final position $M^{(3)}$. The detected positions form a two-dimensional grid whose grid lines are numbered sequentially with respect to their points of intersection (marked with *triangles* for the set of horizontal grid lines and with *squares* for the vertical lines) with two lines x and y (**d**). The *dot* marks the chosen point of intersection of the lines x and y. (**e**) The numbering for grid lines where each of them connects positions which belong to the same {111} plane. (**f**) The resulting indexing for the gridding in the case (**d**)

5. Sequential numbering of the grid lines, separately for each of the two sets of lines (Figs. 4.4d,e), yielding a two-dimensional grid where a pair of indices is assigned to each of the positions.

In the first step, care has to be taken that local brightness maxima between the "main" maxima are excluded. An enhanced intensity between the "main" bright spots of the HRTEM image is often observed under imaging conditions close to those that give "half-spacing" contrast, where the positions of the tunnels and of the rows of atoms show similar brightness. In order to avoid these "artificial" positions, a minimum distance between adjacent positions is defined. If a new maximum position is found during the search procedure, its distances from all other positions found previously are checked. If a distance is smaller than the predefined minimum distance, the

position with the lower intensity is deleted. The detection of pixels with a
local intensity maximum performed in the first step is inaccurate, owing to
residual noise and to the finite size of pixels. Since the intensity profile across
a spot is nearly sinusoidal, it can be fitted by a parabolic curve in a region
close to the estimated intensity maximum. The maximum of the parabolic
curve that is fitted in the second step yields a more accurate estimate of the
peak position, because it is not only the maximum intensity that is used for
the position determination (which can be affected severely by noise). Instead,
a whole manifold of intensity values around the true maximum position is
used, together with an approximate functional relationship. A further gain in
accuracy is achieved in the third step, where parabolas are fitted along four
lines as shown in Fig. 4.4c. The positions of the maxima of the parabolas are
averaged, leading to the final position $M^{(3)}$. For each position, the standard
deviation $\sigma = (\sigma_x, \sigma_y)$ of the positions of the maxima of the four parabolas
is stored. A typical value of $|\sigma|$ is 0.2 pixels.

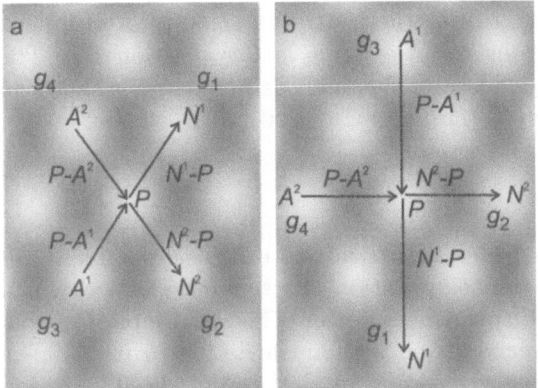

Fig. 4.5. Schematic drawings showing the positions N^1, N^2 and P, and the positions A^1 and A^2. The *arrows* point from a position to the positions of its neighbors. Therefore, only the positions N^1 and N^2 are considered to be neighbors of
P, whereas P is a neighbor of both A^1 and A^2. The parameters $g_{1,2,3,4}$ regulate
the contributions of the individual "distance arrows" to the calculation of local distances. (**a**) corresponds to the grid lines shown in Fig. 4.4e, whereas (**b**) corresponds
to Fig. 4.4d

The generation of grid lines is performed in the third step. The procedure
starts with the selection of two directions $\mathbf{d}_1$ and $\mathbf{d}_2$, along which the grid
lines are intended to run. For each of the detected positions P, its nearest
neighbors N^1 and N^2 in the positive directions of $\mathbf{d}_1$ and $\mathbf{d}_2$ are searched
for (Fig. 4.5). The nearest-neighbor positions in the directions $-\mathbf{d}_1$ and $-\mathbf{d}_2$
are A^1 and A^2, respectively. In this way, strings of neighboring positions are

formed, representing two sets of grid lines, one set (1) with lines along $\mathbf{d}_1$ and the other (2) with lines along $\mathbf{d}_2$.

In the fifth step, a point of the image is chosen that will be used as the intersection point (marked with a dot in Figs. 4.4d,e) of two axes x and y, where the x axis is in the horizontal and the y axis in the vertical direction of the image. The grid lines of sets 1 and 2 intersect the x axis at angles of α_x^1 and α_x^2, respectively, and intersect the y axis at α_y^1 and α_y^2, respectively. If $|\alpha_x^1 - \pi/2| < |\alpha_y^1 - \pi/2|$, then the x axis is used for the indexing of the grid lines of set 1, otherwise the y axis is used. The indexing is performed in such a way that the indices correspond to the order of the intersection points of the grid lines with the x or y axis as appropriate. An analogous procedure is performed with the grid lines of set 2. The result is shown for two different choices of $\mathbf{d}_1$ and $\mathbf{d}_2$: in Fig. 4.4d the [110] and [001] directions have been chosen, and in Fig. 4.4e the two $\langle 111 \rangle$ directions. In this way, we obtain a 2D grid, where each lattice point P is characterized by two indices i and j. Therefore, a lattice point P will be denoted by a spatial vector $\mathbf{P}_{i,j}$ in the following. An example is shown in Fig. 4.4f, which was obtained with the indexing shown in Fig. 4.4d. Note that there may be some positions, such as $i, j = 1, 2$, which do not belong to an existing lattice position $P_{i,j}$. The gridding of the example used to illustrate the analysis procedure is shown in Fig. 4.1, where the directions $\mathbf{d}_1 = [110]$ and $\mathbf{d}_2 = [001]$ were chosen.

4.1.3 Calculation of Lattice Base Vectors

In this section, we describe the generation of lattice base vectors. They will be used in the next section for the computation of a reference lattice, which enables the calculation of local displacement vectors. The lattice base vectors should be deduced from a reference region without deviations from the perfect crystal structure, located far away from any lattice defects. The region indicated by the green rectangle in Fig. 4.1 was chosen as a reference area. The lattice positions inside the reference area are used to calculate two lattice base vectors $\mathbf{a}_1$ and $\mathbf{a}_2$, which correspond to the directions $\mathbf{d}_1$ and $\mathbf{d}_2$. For each grid line, all those positions that lie within the reference region are used to fit a straight line, which results in two sets of straight lines. The directions $\hat{\mathbf{a}}_1$ and $\hat{\mathbf{a}}_2$ of the lattice base vectors are calculated by averaging the gradients of the fitted straight lines. The positions on each grid line found inside the reference area are projected onto the directions $\hat{\mathbf{a}}_1$ and $\hat{\mathbf{a}}_2$. The distances between neighboring projected points are averaged for each of the sets 1 and 2, which yields the lengths of the base vectors $\mathbf{a}_1$ and $\mathbf{a}_2$, respectively. Note that the lattice base vectors obtained in this way should not be understood to constitute lattice translation vectors. In the case shown in Fig. 4.4d, the lattice base vector parallel to the x direction points from the position $\mathbf{P}_{1,1}$ to the (virtual) position $\mathbf{P}_{1,2}$ (Fig. 4.4f). In this case, the grid consists of two sublattices, marked by black and white grid lines in Fig. 4.4f.

4.1.4 Local Displacements and Lattice Spacings

The lattice base vectors $\mathbf{a}_1$ and $\mathbf{a}_2$ obtained in the previous section are now used to calculate reference lattice positions. The positions

$$\mathbf{R}_{i,j} = i\mathbf{a}_1 + j\mathbf{a}_2 - \mathbf{a}_0 \, ,\tag{4.5}$$

which form a reference lattice, can be compared directly with the assiciated positions $\mathbf{P}_{i,j}$. The vector $\mathbf{a}_0$ in (4.5) describes the origin of the reference lattice and is obtained from the condition that the sum of deviations $\mathbf{R}_{i,j} - \mathbf{P}_{i,j}$ calculated inside the reference region (indicated below by "Ref. R.") vanishes:

$$\mathbf{a}_0 = \sum_{i,j:\mathbf{P}_{i,j}\in\text{Ref. R.}} (i\mathbf{a}_1 + j\mathbf{a}_2 - \mathbf{P}_{i,j}) \, .\tag{4.6}$$

The standard deviation of $\mathbf{R}_{i,j} - \mathbf{P}_{i,j}$ computed inside the reference lattice can be used to estimate the accuracy of the position determination. We define

$$\delta = \frac{1}{N^2_{\text{Ref. R.}}} \sqrt{\left[\sum_{\text{Ref. R.}} |\mathbf{R}_{i,j} - \mathbf{P}_{i,j}|\right]^2 - \sum_{\text{Ref. R.}} |\mathbf{R}_{i,j} - \mathbf{P}_{i,j}|^2} \, ,\tag{4.7}$$

where $N_{\text{Ref. R.}}$ is the number of lattice positions inside the reference region. Typical values of δ are of the order of 0.005 nm. In our example, $\delta = 0.004$ nm was obtained. Assuming a crystal lattice parameter of 0.6 nm, we can estimate the accuracy of determination of the local lattice parameter to be 0.5% to 2%.

The next step consists in the definition of local displacement vectors

$$\mathbf{u}'_{i,j} = \mathbf{P}_{i,j} - \mathbf{R}_{i,j} \, .\tag{4.8}$$

For most purposes, the displacement vectors $\mathbf{u}'_{i,j}$ have to be projected onto a certain lattice direction parallel to a selected direction $\mathbf{Q}$ given by

$$\mathbf{Q} = k\mathbf{a}_1 + l\mathbf{a}_2 \, ,\tag{4.9}$$

where the values k and l are small integers that determine the direction onto which the distance vectors will be projected. Using (4.9), we obtain the projected displacements

$$u_{i,j} = \frac{N_{\text{SL}}}{|\mathbf{Q}|} \left(\mathbf{u}'_{i,j} \cdot \frac{\mathbf{Q}}{|\mathbf{Q}|}\right) \, .\tag{4.10}$$

Here N_{SL} is the number of sublattices of the 2D grid; it is equal to 2 in the case of Fig. 4.4f. The choice of the grid line directions shown in Fig. 4.4e would lead to $N_{\text{SL}} = 1$. Note that the normalization with respect to $|\mathbf{Q}|$

yields a projected displacement $u_{i,j} = N_{\text{SL}}$ if $\mathbf{u'}_{i,j} = \mathbf{Q}$. Local lattice spacings between nearest-neighbor positions are defined by

$$\Delta_{i,j} = \frac{N_{\text{SL}}}{|\mathbf{Q}|} \frac{1}{\sum_{m=1}^{4} |g_m|} \left[g_1 \left(\mathbf{N}_{i,j}^1 - \mathbf{P}_{i,j} \right) + g_2 \left(\mathbf{N}_{i,j}^2 - \mathbf{P}_{i,j} \right) \right.$$

$$\left. + g_3 \left(\mathbf{P}_{i,j} - \mathbf{A}_{i,j}^1 \right) + g_4 \left(\mathbf{P}_{i,j} - \mathbf{A}_{i,j}^2 \right) \right] \cdot \frac{\mathbf{Q}}{|\mathbf{Q}|} , \qquad (4.11)$$

where $\mathbf{N}_{i,j}^{1,2}$ and $\mathbf{A}_{i,j}^{1,2}$ have been defined in Sect. 4.1.2. For clarification, $\mathbf{N}_{i,j}^{1,2}$ and $\mathbf{A}_{i,j}^{1,2}$, as well as the positions $\mathbf{P}_{i,j}$, are shown in Figs. 4.5a,b for the most common choices of $\mathbf{d}_1$ and $\mathbf{d}_2$. The values $g_{1,2,3,4} \in \{-1, 0, 1\}$ regulate the contributions of the four distance vectors to $\Delta_{i,j}$.

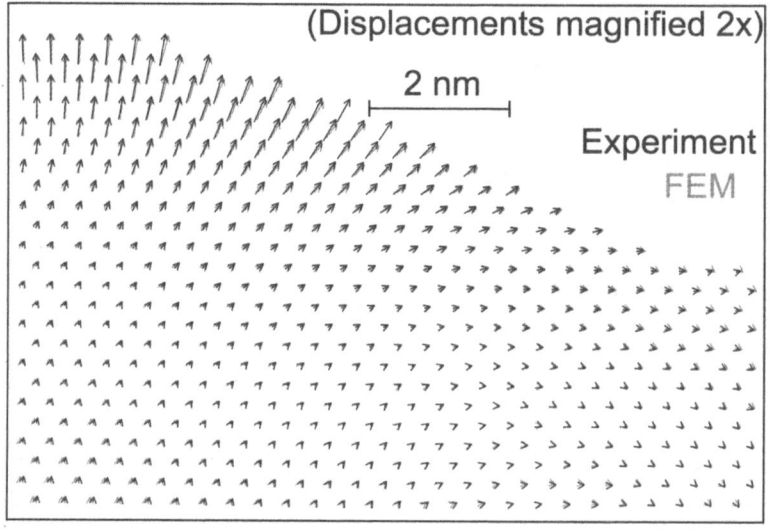

Fig. 4.6. Part of the displacement vector field evaluated from Fig. 4.1 (drawn in *red*) and the corresponding field obtained by FE calculation (*blue*) as outlined in Sect. 4.3.2 [Please see Plate 4 for color reproduction of this plate]

Figure 4.6 depicts the local displacement vectors defined according to (4.8), evaluated with the DALI program package from the right part of the island shown in Fig. 4.1. Owing to the larger lattice parameter in the island, the lengths of the displacement vectors grow from the interface to the top. In the vicinity of the edge of the island, the displacement vectors exhibit a component parallel to the interface plane, which is due to the elastic relaxation of the strained island. The local projected displacements calculated according to (4.10) are displayed in Fig. 4.7 as color-coded maps. In Fig. 4.7a, the displacement vectors have been projected onto the growth direction using $k = -2$, $l = 0$ in (4.9). Figure 4.7b shows the projection of the displacement

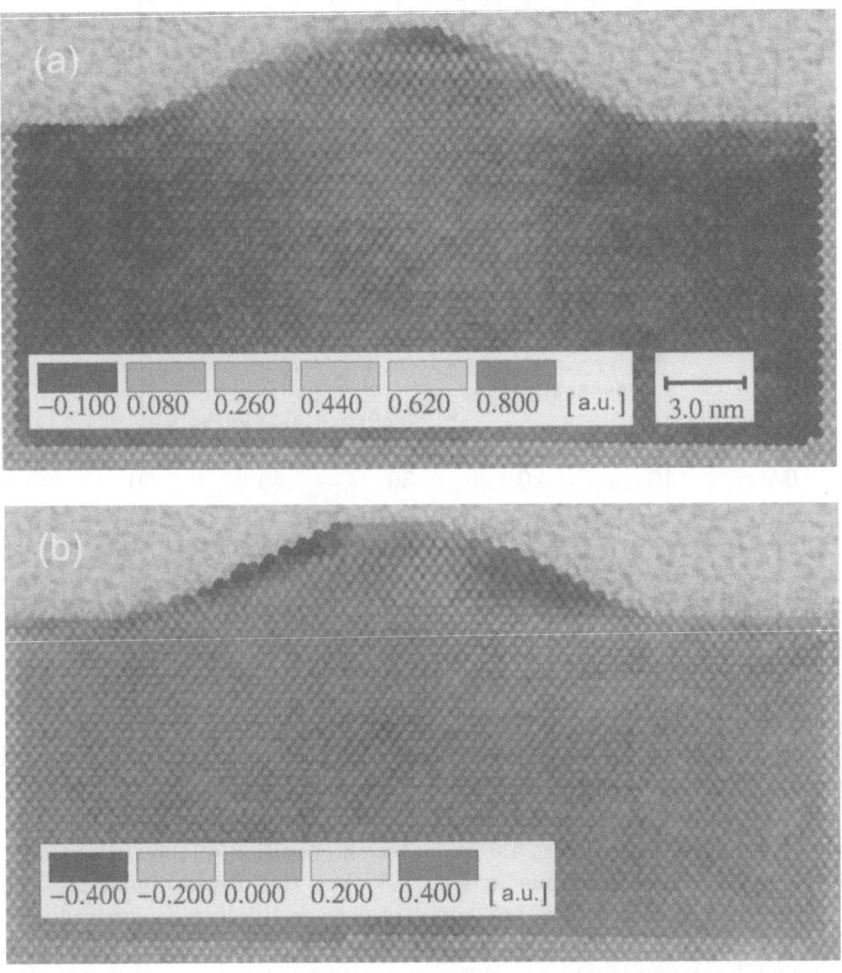

Fig. 4.7. Color-coded maps of the components of the displacement vector field deduced from Fig. 4.1: (**a**) in the growth direction and (**b**) in the interface direction (a positive value indicates a displacement vector pointing to the *right*) [Please see Plate 5 for color reproduction of this plate]

vectors onto the direction parallel to the interface plane, calculated by use of (4.9). The red regions indicate displacement vectors pointing to the right, whereas the blue regions correspond to a component pointing to the left.

4.1.5 Averaged Displacements and Lattice Spacings

In some cases it is appropriate to average the local displacements and lattice spacings along the whole length of a grid line or along part of it. As an example, Fig. 4.8 shows averaged displacements as a function of the number of

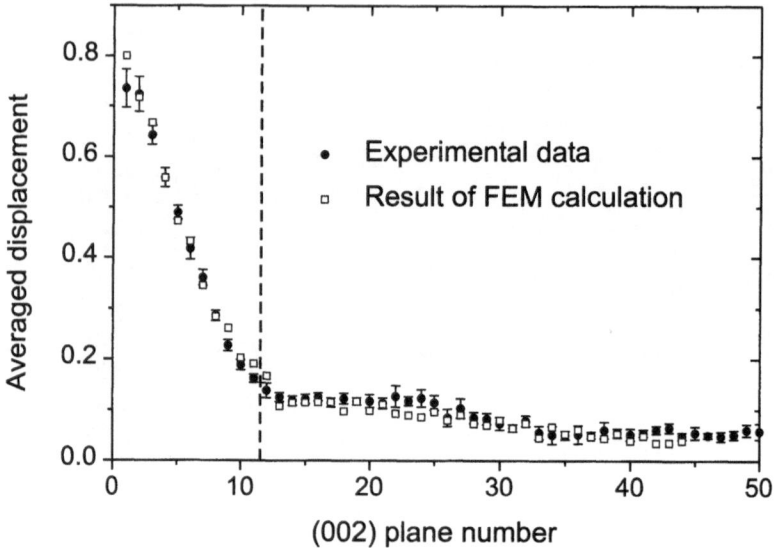

Fig. 4.8. Components of the displacement vector field for each ML in the growth direction along the horizontal direction averaged in the AOI of Fig. 4.1, marked with *filled circles*. The *open squares* are the result of FEM simulations obtained for the FE model with the best fit (see Sect. 4.3.2)

monolayers (ML) number along the growth direction, obtained from the local displacements shown in Fig. 4.7a. The displacements were averaged along the horizontal grid lines inside the AOI marked by a blue rectangle in Fig. 4.1. The vertical dashed line in Fig. 4.8 indicates the position of the surface next to the island. It is conspicuous in Fig. 4.8 that the displacements to the right of the surface marker are not equal to zero as might be expected, but show a weak slope. This effect can be explained by a strain inside the GaAs buffer, which is caused by the biaxially strained island, as was verified by FE simulations as described in Sect. 4.3.2. The preceding discussion clearly shows that a quantitative interpretation of the displacement vector field requires a known correlation between the strain field and the chemical and geometric structure of the sample area investigated. One way to find such a correlation is the application of FE calculations based on a sample geometry derived from the HRTEM image. Whereas the projected shape of the island is directly visible in the HRTEM image, the evaluation of the local sample thickness constitutes a more complicated problem, for which we shall give a possible solution in the next section.

4.2 Determination of the Specimen Thickness

This section deals with the determination of the local sample thickness, which provides a basis for the FE modeling described in Sect. 4.3. Our approach is based on the QUANTITEM procedure, which was suggested by Schwander et al. [10], Ourmazd et al. [11, 12, 13] and Kisielowski et al. [14]. Recently, the QUANTITEM procedure has been also discussed with regard to composition evaluation in ternary mixed crystals such as $In_x Ga_{1-x} As$ [15]. QUANTITEM detects the projected crystal potential, which is proportional to the sample thickness. In [14] it is stated that the QUANTITEM analysis is valid to the extent that dynamical scattering in the investigated material can be described in terms of two Bloch waves. However, it is also shown in [14] that the procedure can be used for III–V semiconductors such as GaAs or AlAs when three Bloch waves are excited with substantial intensity. The present description of the QUANTITEM procedure is based upon the decomposition of the HRTEM micrograph into image unit cells given by the two-dimensional grid whose construction is described in Sect. 4.1. The QUANTITEM method is based on the interpretation of each image unit cell as an image vector. The dimension of the vector is given by the number of pixels included in the cell and should be equal for all cells in the image. The image unit cells obtained by the procedure described in Sect. 4.1 may differ in their sizes and angles. Therefore, the first step of the thickness determination procedure, described in the following section, is a transformation of the image cells that provides square cells of identical sizes $2^n \times 2^n$ pixels (typically $n = 5$).

4.2.1 Cell Transformation

In this section, we describe an algorithm which transforms an irregularly shaped cell Z into a regularly shaped cell Z' (Fig. 4.9b). In order to keep all of the information contained in the cell Z, the size of the cell Z' significantly exceeds the size of the cell Z. To describe the transformation procedure, we define the four vectors $\mathbf{a}, \mathbf{b}, \mathbf{c}$ and $\mathbf{d}$ that point from the center of the area M of the cell Z to the four points at its corners (where M is given by $M_{(x,y)} = \int_Z (x,y)\, \mathrm{d}x\, \mathrm{d}y / \int_Z \mathrm{d}x\, \mathrm{d}y$), and define the vectors $\mathbf{a}', \mathbf{b}', \mathbf{c}'$ and $\mathbf{d}'$ in a similar way for the are M' of the cell Z'.

The following procedure is applied to obtain the intensity of a pixel p'_{nm} inside the regularly shaped cell Z'. For each pixel p'_{nm} its midpoint $\mathbf{p}'_{nm}$ is described by a linear combination

$$\mathbf{p}'_{nm} = \alpha \mathbf{v}'_1 + \beta \mathbf{v}'_2 \tag{4.12}$$

of the two adjacent vectors $\mathbf{v}'_1$ and $\mathbf{v}'_2$ (e.g. $\mathbf{v}'_1 = \mathbf{d}'$ and $\mathbf{v}'_2 = \mathbf{c}'$ for the pixel $\mathbf{p}'_{nm}$ shown in Fig. 4.9b). The corresponding position p in the cell Z is given by

$$\mathbf{p} = \alpha \mathbf{v}_1 + \beta \mathbf{v}_2 \,, \tag{4.13}$$

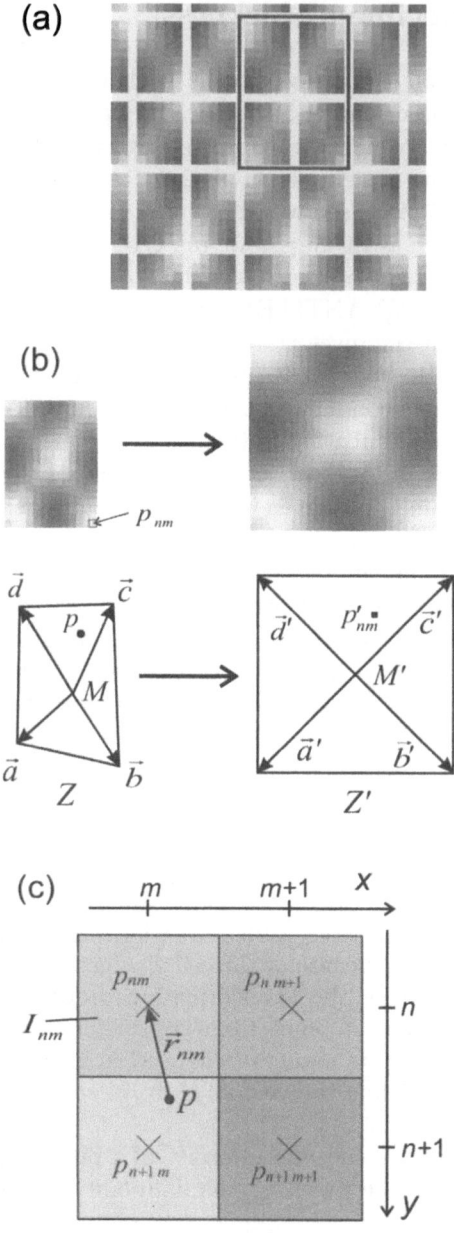

Fig. 4.9. Schematic illustration of the transformation of the original image unit cells into square lattice cells Z'. (**a**) A small part of the HRTEM image shown in Fig. 4.1. The detected lattice sites at the centers of the bright dots are connected by *white lines*. The *black rectangle* indicates the original cell Z. (**b**) The transformation of the cell Z into the square cell Z'. The point p inside the cell Z corresponds to the midpoint of the pixel p'_{nm} inside the cell Z'. (**c**) The system of coordinates used for the description of the transformation procedure. The *crosses* mark the midpoints of the pixels. The brightness of each pixel is described by its intensity value I_{nm}

which generally does not coincide with the midpoint of any pixel in Z. The vectors $\mathbf{v}_1$ and $\mathbf{v}_2$ are the two vectors in the cell Z which correspond to $\mathbf{v'}_1$ and $\mathbf{v'}_2$ in Z' (e.g. $\mathbf{v}_1 = \mathbf{d}$ and $\mathbf{v}_2 = \mathbf{c}$ in Fig. 4.9b). We find four pixels p_{nm}, $p_{n\,m+1}$, $p_{n+1\,m}$ and $p_{n+1\,m+1}$ in Z whose midpoints define a square containing $\mathbf{p}$ (Fig. 4.9c). The values of these four pixels are used to calculate the intensity of the pixel p'_{nm}. For that purpose, four vectors $\mathbf{r}_{nm}$, $\mathbf{r}_{n\,m+1}$, $\mathbf{r}_{n+1\,m}$ and $\mathbf{r}_{n+1\,m+1}$ that point from $\mathbf{p}$ to the midpoints of the four pixels p_{nm}, $p_{n\,m+1}$, $p_{n+1\,m}$ and $p_{n+1\,m+1}$ inside the cell Z (Fig. 4.9c) are defined. Using a coordinate system where the distance between adjacent pixels is 1 (Fig. 4.9c), the intensity I'_{nm} of pixel p'_{nm} is calculated as follows:

$$I'_{nm} = \frac{1}{S} \sum_{i=n}^{n+1} \sum_{j=m}^{m+1} I_{ij} \left(1 - |\mathbf{r}_{ij}|^2 \right) f_{ij} \, ,$$

$$\text{where } S = \sum_{i=n}^{n+1} \sum_{j=m}^{m+1} \left(1 - |\mathbf{r}_{ij}|^2 \right) f_{ij} \text{ and } f_{ij} = \begin{cases} 0 & \text{if } |\mathbf{r}_{ij}| > 1 \\ 1 & \text{otherwise} \end{cases} . \quad (4.14)$$

4.2.2 Determination of Relative Thickness Values

The intensity values I'_{nm} of the pixels p'_{nm} inside a cell Z' are used to define the vector

$$\mathbf{R} = (I'_{11}, \ldots, I'_{1N}, I'_{21}, \ldots, I'_{2N}, \ldots, I'_{NN}) \, , \quad (4.15)$$

where $N = 2^n$ and n is an integer. The QUANTITEM procedure desribed in this section is based on the assumption that the image vectors $\mathbf{R}$ lie (approximately) in a plane E, where they describe an ellipse. The change of the angle of the ellipse corresponds to the change of the projected potential, which is proportional to a change of the sample thickness, for example. In [14] it is shown that this assumption is exactly correct if only two Bloch waves are excited and a small number of beams are used to form the image. However, it is also stated [14] that a wide variety of materials, such as GaAs, can be described in these terms with surprising accuracy.

The plane E is obtained from three template images $\mathbf{R}^{\mathrm{T}}_{1,2,3}$, which are calculated by averaging the image vectors of the cells contained in three small regions. As shown in [14], the result of the QUANTITEM evaluation is nearly independent of the selection of the three (different) regions. The further steps of the QUANTITEM procedure are based on the assumption that each image vector $\mathbf{R}$ can be expressed as a linear combination of the template image vectors

$$\mathbf{R} \cong \rho_1 \mathbf{R}^{\mathrm{T}}_1 + \rho_2 \mathbf{R}^{\mathrm{T}}_2 + \rho_3 \mathbf{R}^{\mathrm{T}}_3 \, , \quad (4.16)$$

which defines a 3D subspace in the N^2-dimensional image vector space. The tips of all image vectors $\mathbf{R}$ lie on a plane E in the 3D image vector space (Fig. 4.10a). The vectors $\widehat{\mathbf{B}}_{1,2}$, given by

(a)

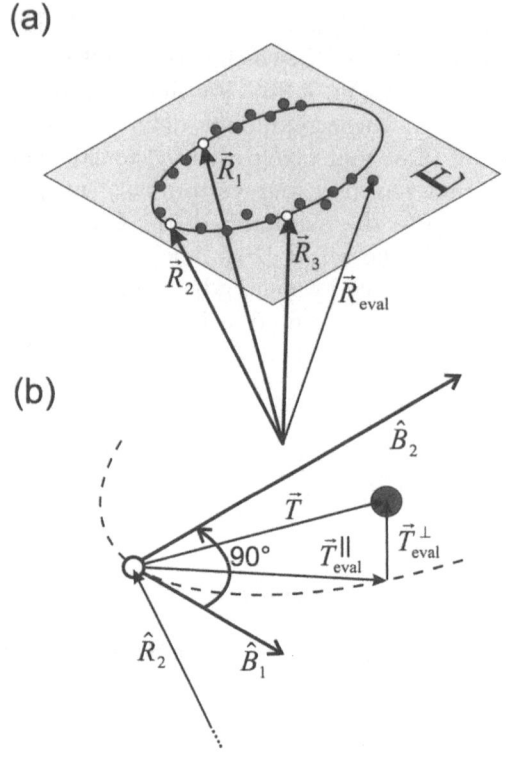

(b)

(c)

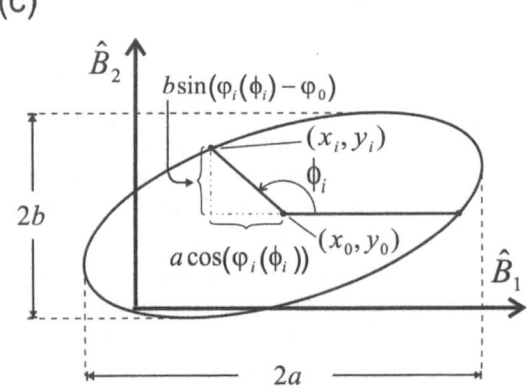

Fig. 4.10. Visualization of the QUANTITEM procedure. (**a**) The plane E, which is defined by three template image vectors $\mathbf{R}_{1,2,3}$. It contains a cloud of experimental data points, represented by vectors (one of them is marked by $\mathbf{R}_{eval}$), whose tips form an ellipse. (**b**) The decomposition of the vector $\mathbf{T}$ into the components $\mathbf{T}^{\parallel}_{eval}$ parallel and $\mathbf{T}^{\perp}_{eval}$ perpendicular to the plane E. $\hat{\mathbf{B}}_1$ and $\hat{\mathbf{B}}_2$ form an orthonormal basis of E. (**c**) Illustration of the meaning of some variables used in 4.19–4.21

$$\widehat{\mathbf{B}}_1 := \frac{\mathbf{R}_1^T - \mathbf{R}_2^T}{|\mathbf{R}_1^T - \mathbf{R}_2^T|} \quad \text{and} \quad \widehat{\mathbf{B}}_2 := \frac{\mathbf{V} - \widehat{\mathbf{B}}_1(\mathbf{V} \cdot \widehat{\mathbf{B}}_1)}{|\mathbf{V} - \widehat{\mathbf{B}}_1(\mathbf{V} \cdot \widehat{\mathbf{B}}_1)|} ,$$

$$\text{where} \quad \mathbf{V} = \frac{\mathbf{R}_3^T - \mathbf{R}_2^T}{|\mathbf{R}_3^T - \mathbf{R}_2^T|} , \tag{4.17}$$

are an orthonormal basis of E (Fig. 4.10b). Owing to the noise in the HRTEM image, the tips of the evaluated image vectors $\mathbf{R}_{\text{eval}}$ may deviate slightly from the plane E. Therefore, we use $\mathbf{R}_{\text{eval}}$ to define an in-plane vector $\mathbf{T}_{\text{eval}}^{\parallel}$ and a vector $\mathbf{T}_{\text{eval}}^{\perp}$ perpendicular to E (Fig. 4.10b), according to

$$\mathbf{T}_{\text{eval}}^{\parallel} = \widehat{\mathbf{B}}_1(\mathbf{T} \cdot \widehat{\mathbf{B}}_1) + \widehat{\mathbf{B}}_2(\mathbf{T} \cdot \widehat{\mathbf{B}}_2) ,$$

$$\text{where} \quad \mathbf{T} = \frac{\mathbf{R}_{\text{eval}}}{|\mathbf{R}_{\text{eval}}|} - \frac{\mathbf{R}_2^T}{|\mathbf{R}_2^T|} \quad \text{and} \quad \mathbf{T}_{\text{eval}}^{\perp} = \mathbf{T} - \mathbf{T}_{\text{eval}}^{\parallel} . \tag{4.18}$$

The value of $\Delta T_{\text{eval}} := |\mathbf{T}_{\text{eval}}^{\perp}|$, which is in the range $\Delta T_{\text{eval}} \cong 0.01 - 0.1$, is used to estimate the reliability of each value of $\mathbf{T}_{\text{eval}}^{\parallel}$. The data $\mathbf{T}_{\text{eval}}^{\parallel} = x_i\widehat{\mathbf{B}}_1 + y_i\widehat{\mathbf{B}}_2$ ($i = 1, 2, \ldots$ numbers the unit cells) obtained for each cell Z_i' are used to fit an ellipse given by $x(\varphi)\widehat{\mathbf{B}}_1 + y(\varphi)\widehat{\mathbf{B}}_2$, where

$$x(\varphi) = a \cos \varphi + x_0 ,$$
$$y(\varphi) = b \sin(\varphi - \varphi_0) + y_0 . \tag{4.19}$$

The values of a, b, x_0, y_0 and φ_0, which are illustrated in Fig. 4.10c, are obtained from a fit procedure. They are used to calculate the angle φ_i corresponding to each cell Z_i' as follows

$$\varphi_i = \arctan\left(\frac{a \sin \omega_i + b \sin \varphi_0 \cos \omega_i}{b \cos \omega_i \cos \varphi_0}\right) ,$$

$$\text{where} \quad \omega_i = \arctan\left(\frac{y_i - y_0}{x_i - x_0}\right) . \tag{4.20}$$

To obtain the sample thickness d_i at the position of each image unit cell, we use the following approximation given in [14]:

$$d_i = \frac{\varphi_i - \Theta}{2\pi} \xi , \tag{4.21}$$

where ξ is the extinction distance of the undiffracted beam along the $\langle 110 \rangle$ ZA (e.g. $\xi = 14.7$ nm for GaAs at 200 kV), and Θ is an angle which defines the origin of the thickness scale. The angle Θ is unknown, and its determination requires some additional information. If the thickness of one unit cell is exactly known, Θ can be obtained from (4.21). The procedure used to obtain Θ is outlined in Sect. 4.2.3.

Another approach, which does not require three template image vectors, uses correspondence analysis (CA) [16]. This procedure is analogous to the interpretation of the cloud of data points given by (4.15) as a distribution of masses that has approximately the shape of a (nearly 2D) ellipsoid. CA is used to find the axes of least inertia, which are the "main axes" of the ellipsoid, by calculation of the eigenvectors and eigenvalues of the matrix of inertia. The two eigenvectors that correspond to the two largest eigenvalues define a plane that is analogous to the plane E shown in Fig. 4.10. The projection of the image vectors onto the plane E yields an ellipse that can be evaluated analogously to the procedure described by (4.19)–(4.21). However, we have found that the application of CA does not provide a significant improvement compared with the use of three image template vectors $\mathbf{R}^{T}_{1,2,3}$. Furthermore, the calculation of the two largest eigenvalues and their corresponding eigenvectors takes about 30 minutes in the case of a 1024×1024 matrix of inertia. Examples where CA has been applied to qualitative analysis of the composition in ternary semiconductor layers may be found in [17, 18].

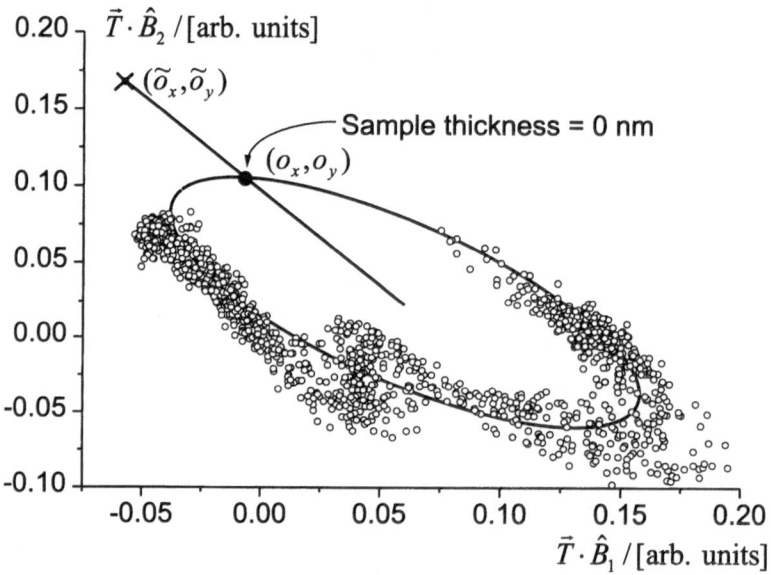

Fig. 4.11. The *open circles* indicate the tips of vectors $\mathbf{T}^{\parallel}_{eval}$ in the plane $(\widehat{\mathbf{B}}_1, \widehat{\mathbf{B}}_2)$. The *solid line* shows an ellipse fitted to the experimental data

We now turn to the evaluation of the experimental HRTEM image shown in Fig. 4.1. However, the results described in this section were derived from a region of the photographic negative that contained a larger part of the substrate (see Fig. 4.13). Figure 4.11 shows the tips of the vectors $\mathbf{T}^{\parallel}_{eval}$ in the $(\widehat{\mathbf{B}}_1, \widehat{\mathbf{B}}_2)$ plane. The solid line shows the ellipse fitted to the experimental data in accordance with (4.19) and (4.20). The thicknesses that correspond to the data points increase in the clockwise direction. Section 4.2.3 outlines two methods to calculate Θ that can be used to determine absolute thickness values.

4.2.3 Determination of Absolute Thickness Values

A simple method to obtain Θ that works with sufficient accuracy in most cases is based upon an image vector $\mathbf{R}_c = (I'_{11}, \ldots, I'_{1N}, I'_{21}, \ldots, I'_{2N}, \ldots, I'_{NN})$, where $I'_{11}, \ldots, I'_{NN} = c$. Here $\mathbf{R}_c$ represents an image cell that contains only one gray level c, corresponding to the image that is expected for a vanishing sample thickness. This cell is added as another data point to the cloud of points formed by use of (4.15), leading to an additional point $(\tilde{o}_x, \tilde{o}_y)$ in E as indicated by a cross in Fig. 4.11. A straight line that connects the center of the ellipse to the point $(\tilde{o}_x, \tilde{o}_y)$ intersects the ellipse at a point (o_x, o_y), which can be regarded as the point of the ellipse that corresponds to a sample thickness of 0 nm. However, in systems where three Bloch waves are excited with substantial intensity, this may constitute an approximation that is not sufficiently accurate. A second procedure for determining the offset angle Θ in (4.21) is based on the method of Stenkamp et al. [19, 20], who suggested consideration of the amplitudes of appropriate Fourier coefficients J_i of the image intensity.

In this second procedure, we consider the amplitude $J(\mathbf{g})$ of the reflection $\mathbf{g}$ in the diffractogram given by (3.14). Its modulus is given by

$$|J(\mathbf{g}, d)| = \sqrt{\Re^2(J(\mathbf{g}, d)) + \Im^2(J(\mathbf{g}, d))} \,, \tag{4.22}$$

where $\Re(z)$ denotes the real part and $\Im(z)$ the imaginary part of a complex number z. The dependence of $J(\mathbf{g})$ on the specimen thickness d in the electron beam direction is indicated explicitly by the use of $J(\mathbf{g}, d)$ in (4.22). The offset angle Θ may now be derived using a known functional relationship for the dependence of $J(\mathbf{g})$ on the specimen thickness. Assuming a sphalerite-type crystal oriented in a $\langle 110 \rangle$ ZA orientation, the relation

$$\lim_{d \to 0} \frac{|J(\mathbf{g}_{111}, d)|}{|J(\mathbf{g}_{000}, d)|} = 0 \tag{4.23}$$

can be used, because the amplitudes of all Bloch waves with $\mathbf{g} \neq 0$ are zero for a vanishing sample thickness. For sufficiently small specimen thicknesses, the following approximation can be used:

$$\frac{|J(\mathbf{g}_{111}, d)|}{|J(\mathbf{g}_{000}, d)|} \propto d \,. \tag{4.24}$$

This works for GaAs in a $\langle 110 \rangle$ projection up to a sample thickness of about 6 nm, as verified by EMS simulations [21].

Figure 4.12 displays $|J(\mathbf{g}_{111}, d)| / |J(\mathbf{g}_{000}, d)|$ evaluated from the part of the experimental HRTEM image previously used for the computation of Fig.

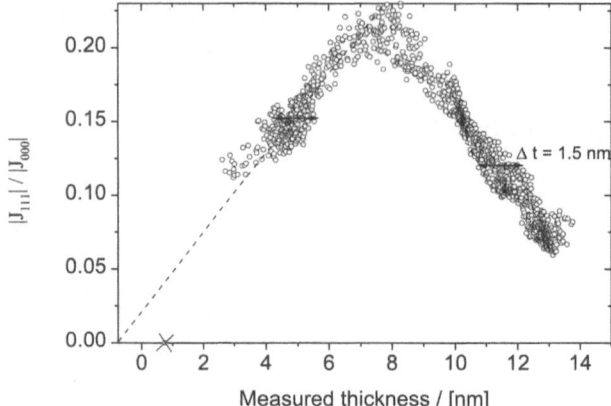

Fig. 4.12. The ratio $|J(\mathbf{g}_{111}, d)|/|J(\mathbf{g}_{000}, d)|$ for each image unit cell plotted versus the measured thickness. The *dashed line* shows the extrapolation of the line fitted to the data below 6 nm. The *cross* on the thickness axis marks the origin of the thickness scale that is obtained using an image unit cell of uniform intensity

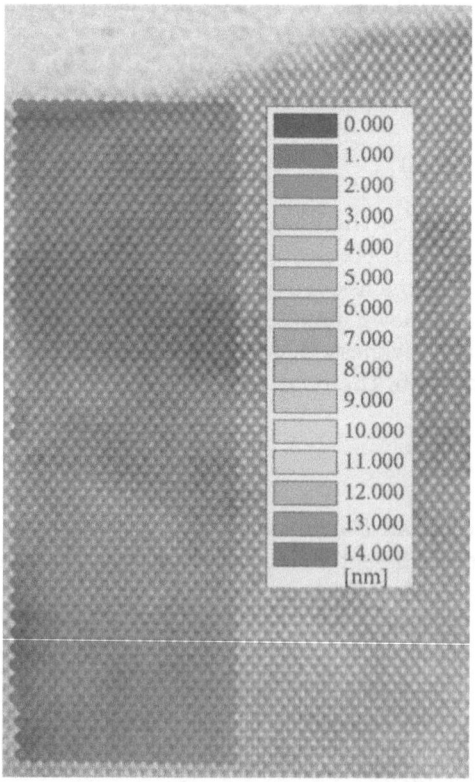

Fig. 4.13. Color-coded map of the evaluated thicknesses [Please see Plate 6 for color reproduction of this plate]

4.11. In this case, the offset angle Θ was calculated as an average of the results obtained by the two methods described above. The first method leads to the origin of the thickness scale that is indicated by a cross in Fig. 4.12. The second method yields an origin that is given by the intersection of the straight line fitted to the data points (marked by a dashed line) below 6 nm with the thickness axis in Fig. 4.12. Taking these result together with the error of the thickness determination obtained from the spread of the data points in the horizontal direction as indicated in Fig. 4.12, we can estimate a maximum error of ± 1.5 nm in the present example. Figure 4.13 shows the resulting thickness map, which reveals a wedge-shaped crystal whose thickness increases from the top to the bottom of the image.

4.3 Elastic Relaxation and Finite-Element Calculation

The determination of the local sample thickness is the basis for the development of a 3D model for the application of the finite-element method, which represents the main subject of the current section. The FE method is used to take account of the elastic relaxation of the tetragonally distorted epitaxial layer, which is due to the small sample thickness in the electron beam direction, typically of less than 20 nm. In the section below, we calculate the tetragonal distortion in the two cases of the thin- and thick-sample limits.

4.3.1 The Analytical Solution
of the Thin- and Thick-Sample Limits

Figure 4.14 depicts the situation for cross-sectional samples of quantum-well-type heterostructures. In both cases the strained layer (gray) is able to expand at the free surfaces of the sample. In the case of the thin sample (Fig. 4.14a), the strained layer is elastically relaxed to the maximum extent. This leads to a diminished tetragonal distortion in comparison with Fig. 4.14b, which corresponds to the case of a very thick sample that is equivalent to the bulk structure. The local lattice parameters or (equivalently) the displacements that are evaluated in Sect. 4.1 depend on the local composition according to Vegard's law, which, for a ternary material $A_x B_{1-x} C$, can be expressed as

$$a_{A_x B_{1-x} C} = x a_{AC} + (1-x) a_{BC} , \tag{4.25}$$

where a_{AC} and a_{BC} are the lattice parameters of the binary components. The main application of strain state analysis is the evaluation of the composition in pseudomorphically grown structures. In this case, the unit cells of the epitaxial layers are tetragonally distorted. The lattice parameter $a_\parallel$ of a unit cell of the layer parallel to the interface plane and perpendicular to the electron beam direction is defined by the lattice parameter a_S of the substrate (Fig. 4.14). The lattice parameter $a_\parallel^{EB}$ parallel to the interface plane and parallel

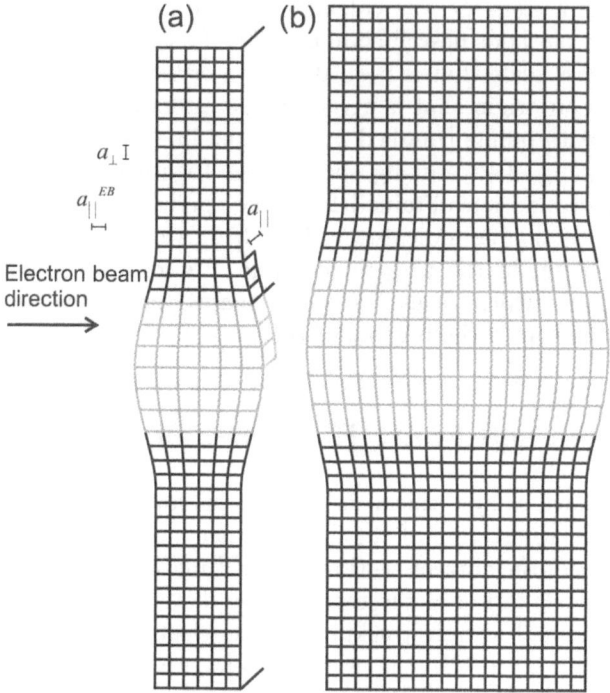

Fig. 4.14. Sketch showing the reduction of the tetragonal distortion of the strained layer (marked in *gray*) in a specimen that is thin in the electron beam direction (**a**), in comparison with a biaxially strained thick sample (**b**)

to the electron beam direction, and also the lattice parameter $a_\perp$ perpendicular to the interface plane, vary locally. For comparison of the FE calculation with the experimental displacement field, we use the approximation that an atomic spacing measured from the HRTEM contrast pattern corresponds to a lattice parameter that is averaged along the electron beam direction. In the following, the lattice parameters $a_\perp$ are calculated under the assumption of a cubic crystal structure, for the electron beam directions $\langle 110 \rangle$ and $\langle 100 \rangle$. If the reference area is chosen inside the substrate (the lower part of the crystal in Fig. 4.14), the lattice parameter that corresponds to the measured lattice spacing in the growth direction (see (4.11)) is in the two limiting cases of a thin or a thick sample, given by

$$\frac{a_S - a_\perp}{a_S} = \alpha_R \frac{a_S - a}{a_S} , \tag{4.26}$$

$$\text{where} \quad [\alpha_R]^{\text{thick}}_{\langle 100 \rangle, \langle 110 \rangle} = \left(1 + 2\frac{C_{12}}{C_{11}}\right) \approx 2 ,$$

$$[\alpha_R]^{\text{thin}}_{\langle 100 \rangle} = \left(1 + \frac{C_{12}}{C_{11} + C_{12}}\right) \approx 1.3 ,$$

$$[\alpha_R]^{\text{thin}}_{\langle 110 \rangle} = \left(1 + 4\frac{C_{44}C_{12}}{C_{11}^2 + 2C_{11}C_{44} + C_{11}C_{12} - 2C_{12}^2}\right) \approx 1.5 .$$

Here a is the (local) bulk lattice parameter and C_{ij} are the elastic constants of the strained layer. Table 4.1 gives an overview of the elastic constants and lattice parameters of the semiconductors considered in this book. Equation (4.26) can be used for the calculation of the error in the composition determination if the sample thickness is not known. An analytical solution analogous to (4.26) for any sample thickness is given in [22].

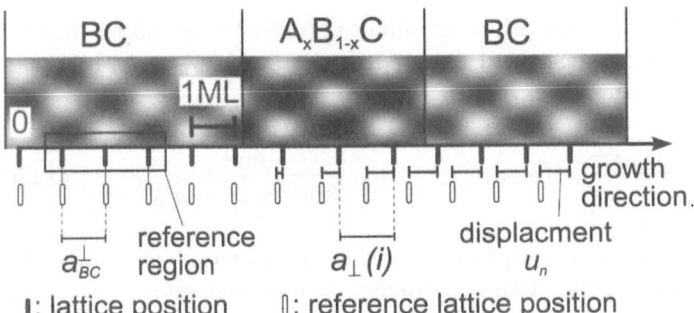

Fig. 4.15. Schematic drawing explaining the increasing displacements $\mathbf{u}_n$ in a region of a ternary material $A_x B_{1-x} C$ with a larger lattice parameter than that of the binary compound BC if the reference region is chosen inside the material BC

Table 4.1. Elastic constants and lattice parameters of semiconductors considered in this book

	Material			
	GaAs	InAs	ZnSe	CdSe
Lattice parameter (nm)	0.5654	0.6058	0.5669	0.6081
C_{11} (GPa)	118.1	83.3	81.0	66.7
C_{12} (GPa)	53.2	45.3	48.8	46.3
C_{44} (GPa)	59.4	39.6	44.1	22.3

Assuming one of the extreme sample thicknesses considered above, we can also obtain an approximation for the measured displacements. From (4.8) and (4.10), we obtain the following for the (averaged) displacement of the grid line n parallel to the interface plane (Fig. 4.15):

$$u'^{\perp}_n = \frac{1}{a_S} \sum_{i=1}^{n} (a_{\perp}(i) - a_S)$$

$$= -\sum_{i=1}^{n} \frac{a_S - a_{\perp}(i)}{a_S} = -\alpha_R \sum_{i=1}^{n} \frac{a_S - a(i)}{a_S} \ . \tag{4.27}$$

If the strained layer is a ternary material, we define the term "integral" AC content in the binary compound C_{AC} of a ternary material $A_xB_{1-x}C$ in the layer as

$$C_{AC} = \sum_{i=1}^{n} x(i) \quad \text{in units of ML AC} \ , \tag{4.28}$$

where the distance between the adjacent planes i and $i+1$ of grid lines parallel to the interface plane is designated as one ML. From (4.25) and (4.27), we deduce, assuming $a_S = a_{BC}$,

$$u_n = \alpha_R \frac{a_{AC} - a_{BC}}{a_{BC}} \sum_{i=1}^{n} x(i) \Rightarrow C_{AC} = u_{\max} \frac{a_{BC}}{\alpha_R(a_{AC} - a_{BC})} \ , \tag{4.29}$$

where $u_{\max}$ is the maximum displacement that is measured on top of the strained layer and a_{BC} is the lattice parameter in the substrate.

4.3.2 Finite-Element Calculations

In the previous section it was shown that the lattice mismatch of a strained layer causes a tetragonal distortion, which is reduced by a small specimen thickness. A further elastic relaxation can take place in an island which is able to expand at its free surfaces. In the case of composition evaluation of islands, an FE calculation is recommended. An FE calculation is also advisable for the investigation of two-dimensional buried or free-standing layers if the accuracy of a composition determination based on (4.26) is not sufficient. The FE calculation starts with the generation of a 3D geometric model of the specimen, which we have performed with the MSC-PATRAN program [23]. For this purpose, we use the projected shape of the sample visible in the HRTEM micrograph and the local sample thickness determined according to Sect. 4.2. The 3D model is composed of "solids", each with a uniform composition. An island or a 2D layer is therefore represented by a stack of solids, where the thicknesses of the solids in the growth direction are usually of the order of one to four MLs. Figure 4.16 shows the decomposition

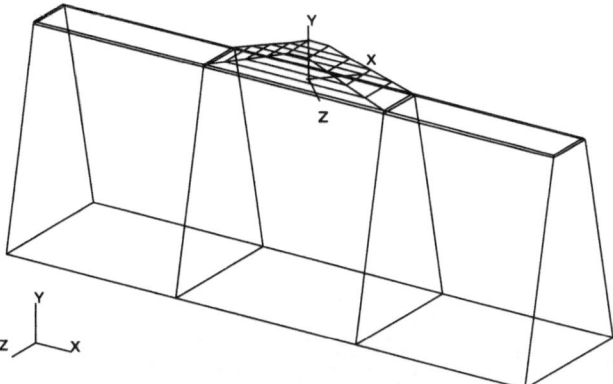

Fig. 4.16. Geometry of the FE model and its decomposition into solids of uniform composition. The *lower* coordinate system is used for the definition of the geometry and the boundary conditions, whereas the *upper* system is associated with the crystallographic notation

of the FE model into solids in the case of the example considered in this chapter.

In order to simulate an assumed concentration profile, elastic constants, a virtual thermal expansion coefficient α_{thermal} and a heating temperature ΔT are assigned to each solid, with the appropriate material parameters. The expansion coefficients are introduced to simulate, by a temperature rise ΔT, the local lattice mismatch that will occur during the FE calculation. In practice, $\Delta T = 0$ is chosen for the solids of the substrate and $\Delta T = 1$ for the solids of the strained material. Therefore, the thermal expansion coefficient of a solid has to fulfill the condition

$$a_S \left(1 + \alpha_{\text{thermal}} \Delta T\right) = a \;\Rightarrow\; \alpha_{\text{thermal}} \Delta T = -\frac{a_S - a}{a_S} \,, \tag{4.30}$$

where a_S is the lattice parameter of the substrate and a the bulk lattice parameter corresponding to the material of the solid.

One also has to define a coordinate system $(\widehat{\mathbf{x}}_{\text{ec}}, \widehat{\mathbf{y}}_{\text{ec}}, \widehat{\mathbf{z}}_{\text{ec}})$ associated with the elastic constants, which may deviate from the orientation of the coordinate axes $(\widehat{\mathbf{x}}_{\text{geom}}, \widehat{\mathbf{y}}_{\text{geom}}, \widehat{\mathbf{z}}_{\text{geom}})$ used for the generation of the geometric model. For the generation of the geometry, the $\widehat{\mathbf{y}}_{\text{geom}}$ axis is usually chosen parallel to the growth direction and the $\widehat{\mathbf{z}}_{\text{geom}}$ axis parallel to the electron beam direction. If the latter is a crystallographic $\langle 110 \rangle$ direction, the coordinate system associated with the elastic constants can be formulated as $(\widehat{\mathbf{x}}_{\text{ec}}, \widehat{\mathbf{y}}_{\text{ec}}, \widehat{\mathbf{z}}_{\text{ec}}) = (\widehat{\mathbf{x}}_{\text{geom}} + \widehat{\mathbf{z}}_{\text{geom}}, \widehat{\mathbf{y}}_{\text{geom}}, -\widehat{\mathbf{x}}_{\text{geom}} + \widehat{\mathbf{z}}_{\text{geom}})$. Boundary conditions have to be defined for the displacements $\mathbf{u}$ of the bottom plane of the FE model $(\mathbf{u}_{x\,\text{geom}} = \mathbf{u}_{y\,\text{geom}} = \mathbf{u}_{z\,\text{geom}} = 0)$ and for the side planes $(\mathbf{u}_{x\,\text{geom}} = 0)$.

The third step is the decomposition of the solids into finite elements, where care has to be taken that the element density is high in regions where

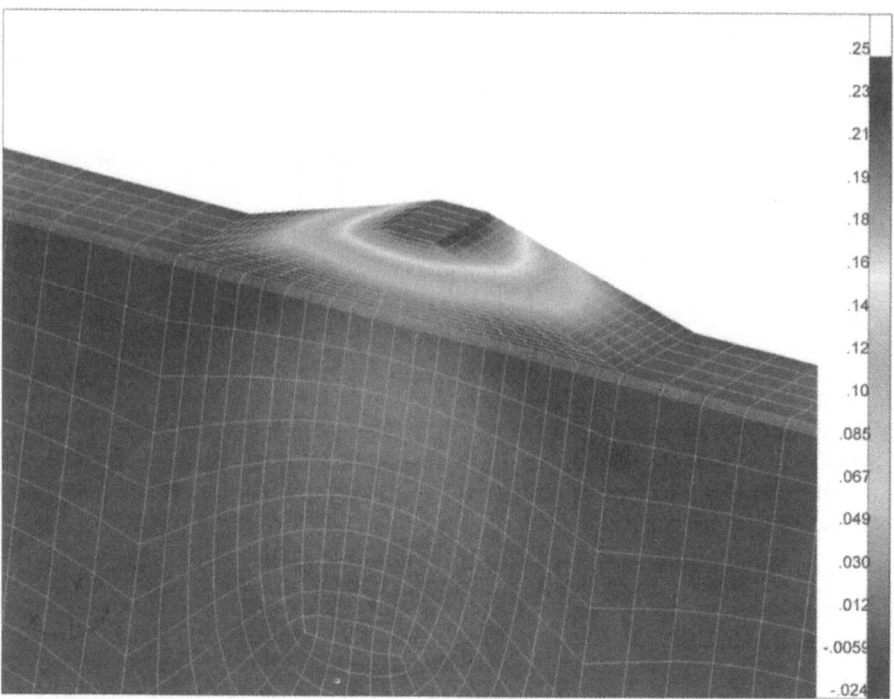

Fig. 4.17. FE model with color-coded values of the components of the displacement vectors in the growth direction. The color-coded scale is given in nanometers. The *light-blue grid* indicates the finite elements [Please see Plate 7 for color reproduction of this plate]

large variations in the displacement are expected (e.g. inside an island or a 2D strained layer). Figure 4.17 shows the decomposition of the FE model into finite elements for the example considered in this chapter. In our case, the structural data were written to a file, which was an input file for the ABAQUS solver [24]. Figure 4.17 also depicts a color-coded map of the resulting displacements in the growth direction, given in nanometers. We have applied the following steps to the result of the FE modeling to obtain a direct comparability with the experimental displacement values determined by strain state analysis of the HRTEM image. In the first step, atomic positions in the three-dimensional FE model of the specimen are calculated; these depend on the crystal structure and orientation. Next, the atomic displacements are determined by interpolation of the surrounding nodal displacements. Finally, the atomic positions and displacements are averaged along the atomic rows in the electron beam direction. As a result, a 2D field of projected atomic positions and displacements is obtained, which can be evaluated with the DALI program. The result of the FEM simulation is then compared with the

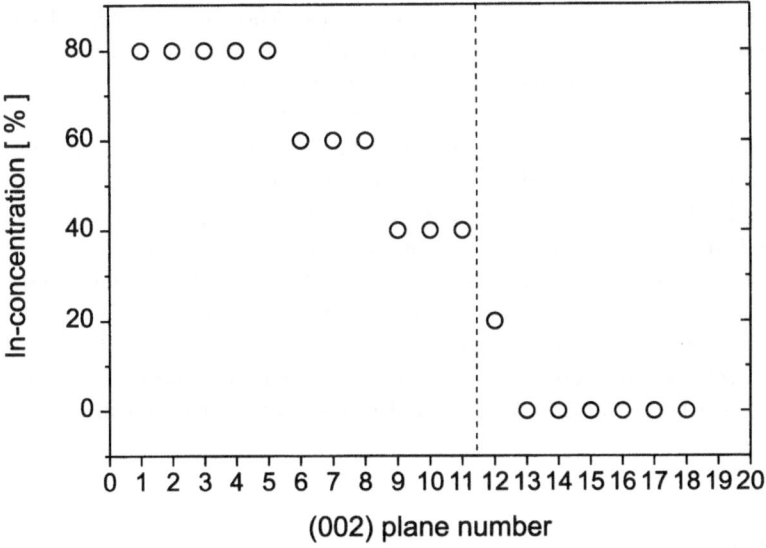

Fig. 4.18. Indium concentration obtained from the analysis (see text) plotted versus the (002) plane number. The *dashed line* marks the position of the surface next to the island. The finite In content to the *right-hand side* of the dashed line corresponds to a wetting layer with a thickness of one ML

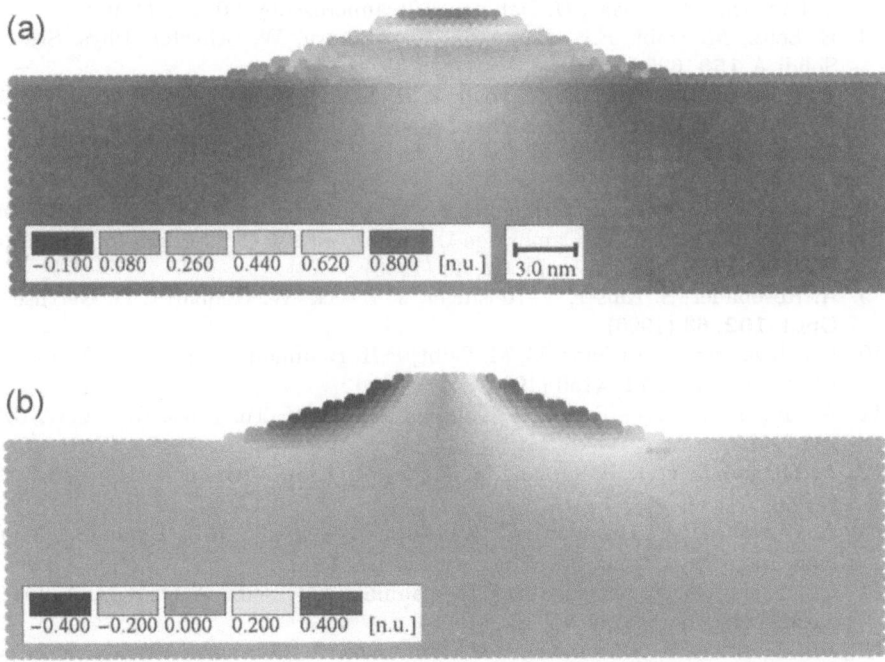

Fig. 4.19. Components of the displacement vector field in (**a**) the growth direction and (**b**) the horizontal direction, evaluated from the FE calculation [Please see Plate 8 for color reproduction of this plate]

experimental displacements. In an iterative process, the compositions of the solids are changed until sufficient agreement with experiment is achieved.

The displacement vectors that result from the FE calculation (blue in Fig. 4.6) can be compared with the experimental displacement vectors (red in the same figure) evaluated from part of the HRTEM image. The high degree of coincidence shows the validity of the FEM approach to the nanoscopic problem. Figure 4.18 displays the In concentration profile in the growth direction used in the FE model. Note that the assumed In concentration does not vary along the planes parallel to the interface. Figure 4.19 shows color-coded maps of (a) the displacements in the growth direction and (b) the displacements in the [110] direction. The scaling has been chosen to be identical to that of Fig. 4.7 to reveal the good agreement between Figs. 4.19 and 4.7. Figure 4.8 contains the FE displacements averaged in a region corresponding to the AOI in Fig. 4.1.

References

1. R. Bierwolf, M. Hohenstein, F. Phillipp, O. Brandt, G.E. Crook, K. Ploog: Ultramicroscopy **49**, 273 (1993)
2. O. Brandt, K. Ploog, R. Bierwolf, M. Hohenstein: Phys. Rev. Lett. **68**, 1339 (1992)
3. S. Paciornik, R. Kilaas, U. Dahmen: Ultramicroscopy **50**, 255 (1993)
4. H. Seitz, M. Seibt, F.H. Baumann, K. Ahlborn, W. Schröter: Phys. Status Solidi A **150**, 625 (1995)
5. P.H. Jouneau, A. Tardot, G. Feulliet, H. Marietta, J. Cibert: J. Appl. Phys. **75**, 7310 (1994)
6. T. Schuhrke, M. Mändl, J. Zweck, H. Hoffmann: Ultramicroscopy **45**, 411 (1992)
7. W.H. Press, W.T. Vetterling, S.A. Teukolsky, B.P. Flannery: *Numerical Recipes in C*, 1st edn. (Cambridge University Press, Cambridge, 1992)
8. L.D. Marks: Ultramicroscopy **62**, 43 (1996)
9. A. Rosenauer, S. Kaiser, T. Reisinger, J. Zweck, W. Gebhardt, D. Gerthsen: Optik **102**, 63 (1996)
10. P. Schwander, C. Kisielowski, M. Seibt, F.H. Baumann, Y. Kim, A. Ourmazd: Phys. Rev. Lett. **71**, 4150 (1993)
11. A. Ourmazd, D.W. Taylor, J. Cunningham, C.W. Tu: Phys. Rev. Lett. **62**, 933 (1989)
12. A. Ourmazd, F.H. Baumann, M. Bode, Y. Kim: Ultramicroscopy **34**, 237 (1990)
13. A. Ourmazd, P. Schwander, C. Kisielowski, M. Seibt, F.H. Baumann, Y.O. Kim: Inst. Phys. Conf. Ser. **134**: Section 1, 1 (1993)
14. C. Kisielowski, P. Schwander, F.H. Baumann, M. Seibt, Y.O. Kim, A. Ourmazd: Ultramicroscopy **58**, 131 (1995)
15. J.L. Maurice, P. Schwander, F.H. Baumann, A. Ourmazd: Ultramicroscopy **68**, 149 (1997)
16. J.F. Aebersold, P.A. Stadelmann, J.-L. Rouvière: Ultramicroscopy **62**, 171 (1996)

17. K.G. Chinyama, K.P. O'Donnell, A. Rosenauer, D. Gerthsen: J. Cryst. Growth **203**, 362 (1999)
18. T. Walter, A. Rosenauer, R. Wittmann, D. Gerthsen, F. Fischer, T. Gerhard, A. Waag, G. Landwehr, P. Schunk, T. Schimmel: Phys. Rev. B **59**, 8114 (1999)
19. D. Stenkamp, W. Jäger: Inst. Phys. Conf. Ser. **134**: Section 1, 15 (1993)
20. D. Stenkamp, H.P. Strunk: Appl. Phys. A **62**, 369 (1996)
21. P.A. Stadelmann: Ultramicroscopy **51**, 131 (1987)
22. M.M.J. Treacy, J.M. Gibson: J. Vac. Sci. Technol. B **4**, 1458 (1986)
23. PATRAN 7.0, (MacNeal-Schwendler Corporation, Los Angeles)
24. ABAQUS 5.5, (Hibbitt, Karlsson & Sorenson Inc., Rhode Island)

5 Lattice Fringe Analysis

An alternative composition evaluation procedure is presented in this chapter which does not exploit the information contained in lattice parameter fluctuations but is based upon a chemically sensitive reflection. Therefore, it may be regarded as an analysis method that is complementary to the strain state analysis described in Chap. 4. Chemically sensitive reflections such as the {002} reflection in the sphalerite-type crystal structure have already been introduced in Sect. 2.2 (see Fig. 2.7). This type of reflection depends strongly on the chemical composition and, therefore, is particularly well suited for compositional analysis.

This chapter is organized as follows. First, we make give some basic definitions that will help us to distinguish clearly between reflections in the diffractogram and beams in the diffraction pattern, and we define their amplitudes, moduli and phases. In the approach described here, the images are evaluated locally, which needs a definition of what local values of beam (or reflection) amplitudes are supposed to mean. We then explain some of the considerations which led to the development of the CELFA method. In this context, we aim at finding optimized imaging conditions, which are assumed in subsequent sections. The theoretical treatment starts with a consideration of the beams that contribute to the image formation, and is followed by an analytical calculation of the amplitudes of reflections in the diffractogram. The next topic is the estimation of the specimen thickness based upon a three-beam condition and the exposure of a defocus series. After presenting an example of such an evaluation, we shall be concerned with some more practical aspects that influence the accuracy of the measurement.

5.1 On the Basic Ideas Behind CELFA

5.1.1 General Aspects and Definitions

For notational convenience, we consider first the chemically sensitive (002) beam of the the sphalerite-type crystal structure. Later we shall discuss the differences between those chemically sensitive {002} beams which do not appear to be equivalent in pseudomorphically grown strained-layer heterostructures. The present chapter on the CELFA technique is mainly concerned with the question of *how to measure and evaluate the **local amplitude** of the chemically sensitive (002) **beam** with the highest possible accuracy.*

For clarity, we define the term *beam* here such that it is related to the Fourier transform of the wave function at the object exit surface (diffraction pattern), whereas we use the term *reflection* when referring to the Fourier transform of the image intensity (diffractogram). The expression *local amplitude of the beam* g also needs some explanatory definitions.

Local amplitude of a beam. First of all, note that "local" refers not to Fourier space but to real space. Now consider the object exit wave in the vacuum in a plane close to the surface of the specimen at a position r. Cut out a patch of size $\Delta x_1 \times \Delta x_2$ centered around r where, for example, $\Delta x_1 = \Delta x_2 = 1/g$ (x_1 and x_2 are spatial coordinates in the plane considered) such that the periodic continuation of the patch is smooth in both if the directions $\hat{x}_1$ and $\hat{x}_2$. Now generate a new wave function with infinite extent along $\hat{x}_1$ and $\hat{x}_2$ by periodic continuation of the cut-out patch. Apply a Fourier transform. The resulting amplitude of the Fourier coefficient corresponding to the spatial frequency g (in the sense described by (3.9)) is the *local amplitude* $F(g, r)$ of the beam g. In the following considerations, it is also important to note that $F(g, r)$ is a complex-valued quantity, which can be written as

$$F(\mathbf{g}, \mathbf{r}) = a(\mathbf{g}, \mathbf{r}) \exp\{ip(\mathbf{g}, \mathbf{r})\} . \tag{5.1}$$

This contains the *local modulus of the beam* g, given by $a(\mathbf{g}, \mathbf{r}) = |F(\mathbf{g}, \mathbf{r})|$, and the *local phase of the beam* g, given by $p(\mathbf{g}, \mathbf{r})$. Owing to the small scattering angles that occur in high-energy electron scattering, the local amplitude $F(\mathbf{g}, \mathbf{r})$ contains, to a very good approximation, information about the specimen at the position r; this is analogous with the "column approximation" [1] frequently applied in TEM.

The requirement for good accuracy of the measured local modulus of the (002) beam implies a good signal-to-noise ratio, a small dependence on specimen thickness fluctuations and a small dependence on local variations of imaging parameters such as the defocus. Note that we define the expression *local defocus* here as the local distance between the lower surface of the specimen and the object plane defined by the excitation of the objective lens. Variations of the local defocus may occur if the thickness of the specimen is not constant laterally.

The necessity of knowing the local phase of the (002) beam can easily be understood by considering the dependence of the structure factor on the composition, as shown for $In_x Ga_{1-x} As$ in Fig. 2.7a. It is obvious that the structure factor is negative for an In concentration x smaller than $x_0 \approx 22\%$ and positive for $x > x_0$. Consider two regions of the specimen with $x = x_0 + \Delta x$ and $x = x_0 - \Delta x$, respectively. Figure 2.7a shows that the structure factor depends linearly on x. The moduli of the (002) beam are consequently equal in the two regions considered. To get rid of this ambiguity, the phase of the (002) beam has to be taken into account; this phase has a difference of π between the regions with $x = x_0 + \Delta x$ and $x = x_0 - \Delta x$. Therefore, it is desirable to know at least the *sign* $s(\mathbf{g}_{002}, x)$ of the (002) beam. For clarity, we define the sign of a beam $\mathbf{g}$ in accordance with its structure factor:

$$s(\mathbf{g}, x) := \operatorname{sign}\left(F_S(\mathbf{g}, x)\right),\tag{5.2}$$

where F_S is the structure factor defined in (2.33). The sign of the (002) beam for $In_x Ga_{1-x} As$ is given by $s(\mathbf{g}_{002}, x) = -1$ for $x < x_0$ and $s(\mathbf{g}_{002}, x) = +1$ for $x > x_0$, where x_0 is the In concentration at which the structure factor is zero. For convenience, we also define the *local signed modulus of the beam* $\mathbf{g}$ by

$$a_S(\mathbf{g}, \mathbf{r}) = s(\mathbf{g}, x(\mathbf{r}))|F(\mathbf{g}, \mathbf{r})| \, .\tag{5.3}$$

Local amplitude of a reflection. In the following, we give analogous definitions for the reflections in the diffractogram. First, the *local amplitude of the reflection* $\mathbf{g}$ is given by

$$J(\mathbf{g}, \mathbf{r}) = A(\mathbf{g}, \mathbf{r}) \exp\{iP(\mathbf{g}, \mathbf{r})\} \, ,\tag{5.4}$$

which defines the *local modulus* $A(\mathbf{g}, \mathbf{r})$ and the *local phase* $P(\mathbf{g}, \mathbf{r})$ of the reflection $\mathbf{g}$. The *local signed modulus* $A_S(\mathbf{g}, \mathbf{r})$ of the reflection $\mathbf{g}$ is given, in analogy with (5.3), by

$$A_S(\mathbf{g}, \mathbf{r}) = s(\mathbf{g}, x(\mathbf{r}))|J(\mathbf{g}, \mathbf{r})| \, .\tag{5.5}$$

The local amplitudes $J(\mathbf{g}, \mathbf{r})$ are related to the image diffractogram by

$$I(\mathbf{r}) = \sum_{\mathbf{g}} J(\mathbf{g}, \mathbf{r}) \exp\{-2\pi i \mathbf{g} \cdot \mathbf{r}\} \, .\tag{5.6}$$

We still need the definitions of the nonlocal values of amplitudes and phases, etc. To avoid further indexing or labeling, we shall use the same expressions for local and nonlocal values; however, we shall leave out the dependence on $\mathbf{r}$ and replace it by other dependences, e.g. a concentration dependence. For example, we would represent the concentration-dependent modulus of the (002) beam found by Bloch wave calculations by $a(\mathbf{g}_{002}, x)$, whereas $a(\mathbf{g}_{002}, \mathbf{r})$ could be used for the modulus evaluated from the image at the position $\mathbf{r}$. The particular meaning intended should be clear from the context.

Summarizing the considerations in the preceding paragraphs, we can say that the problem of measuring the local composition using the chemically sensitive (002) reflection may be solved by the following idealized method:

1. Measurement of the local values of the modulus and sign of the (002) beam.
2. Measurement of the local specimen thickness.

3. Computation of the local composition by comparison of the measurement with the modulus and sign of the (002) beam obtained by Bloch wave calculations.

In the next section, we discuss how this idealized concept could be applied in practice.

5.1.2 Discussion of Optimum Imaging Conditions

5.1.2.1 Impact of Nonlinear Image Formation Let us start the discussion with the first point of the list above. It was pointed out in Chap. 3 that the modulus and the phase of a beam are not directly accessible in the TEM ,where we record the local intensity of the electron wave in the image plane. The Fourier transform of the image intensity (the diffractogram) can be used to obtain a relationship between the beams contributing to the image. It was also pointed out in Chap. 3 that a diffractogram reflection corresponding to a spatial frequency $\mathbf{g}_p$ is not related to only one beam, but contains contributions from interferences of pairs of beams $(\mathbf{g}_n, \mathbf{g}_m)$ that obey $\mathbf{g}_p = \mathbf{g}_n - \mathbf{g}_m$.

For a sufficiently thin specimen, the undiffracted beam $\mathbf{g} = 0$ has the largest amplitude. If we consider the reflection $\mathbf{g}_p$ of the diffractogram, the linear contribution that stems from the interference of the undiffracted beam with the beam $\mathbf{g}_p$ is (usually) dominant. Consequently, the (002) reflection of the diffractogram is of special interest here because its linear contribution is strong and proportional to the chemically sensitive (002) beam.

5.1.2.2 Conventional Dark-Field Imaging The above considerations reveal the disadvantages of conventional dark-field (DF) imaging, where, in our case, only the (002) reflection would pass through the objective lens aperture. An example is given in Fig. 5.1a, which shows a DF image of an $In_xGa_{1-x}As/GaAs(001)$ SK layer (marked with a black arrow) covered with 10 nm GaAs. Figure 5.1b displays the relationship between the (normalized) image intensity and the In concentration for various specimen thicknesses up to 150 nm. In Fig. 5.1a, the bright regions are the centers of In-rich islands. The maximum In concentration is about 60%. The corresponding diffractogram contains only a reflection at $\mathbf{g} = 0$, which corresponds to the (002) beam that was centered on the optical axis to obtain the image. The local amplitude of this beam $J(\mathbf{g}_{000}, \mathbf{r}) = F(\mathbf{g}_{002}, \mathbf{r})^2$ is usually small. An example can be seen in Fig. 2.9a for GaAs, where $F(\mathbf{g}_{002}, \mathbf{r})^2 < 0.01$, the amplitude being normalized with respect to $F(\mathbf{g}_{000}) = 1$ for an infinitely small specimen thickness. As a consequence, a measurement of $J(\mathbf{g}_{000}, \mathbf{r})$ obtained from a (002) DF image has a small signal-to-noise ratio. Using a linear interference instead boosts the signal-to-noise ratio by a factor of ≈ 10, because $J(\mathbf{g}_{002}, \mathbf{r}) = F(\mathbf{g}_{002}, \mathbf{r})F(\mathbf{g}_{000}, \mathbf{r}) \lesssim 0.1$ is about ten times larger than $F(\mathbf{g}_{002}, \mathbf{r})^2$.

Additionally, the phase of the (002) beam is lost in an (002) DF image. This means that the ambiguity of the modulus of the (002) reflection

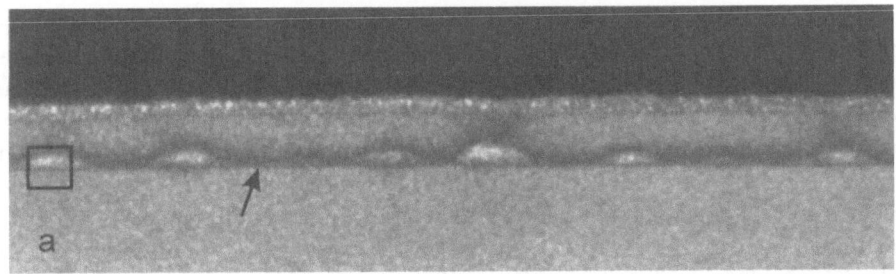

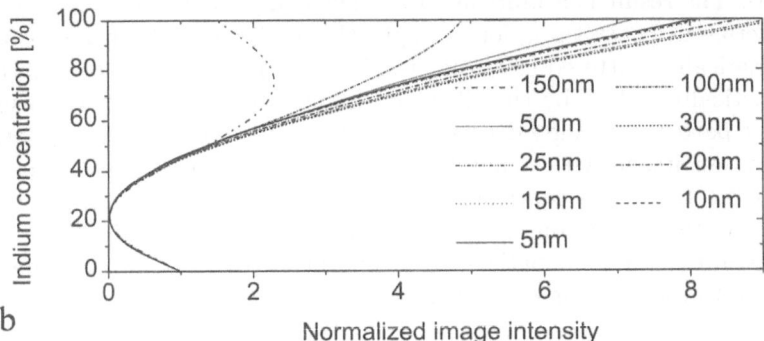

Fig. 5.1. (a) $\langle 100 \rangle$ TEM DF image of an $In_xGa_{1-x}As/GaAs(001)$ SK layer capped with 10 nm GaAs, obtained with a strongly excited (002) beam centered on the optical axis. (b) Indium concentration plotted versus the normalized image intensity, calculated for an image obtained with the (002) beam centered on the optic axis. The curves corresponding to different sample thicknesses, as given in the legend, were calculated by the Bloch wave method with the EMS program package. Each computed curve is normalized with respect to the intensity in the GaAs $(x = 0)$ at the appropriate thickness

discussed in the previous section cannot be resolved. Regions with concentrations $x = x_0 + \Delta x$ and $x = x_0 - \Delta x$ would thus yield the same measured quantity $J(\mathbf{g}_{000}, \mathbf{r})$.

A third disadvantage of using (002) DF images is the larger impact of inelastically scattered electrons. This effect can also be seen in Fig. 5.1a. Consider the black frame at the left side of the image. The In concentration increases from the bottom of the frame towards the center of the island, from 0 to 60%. The dark line at the bottom of the island is due to the intensity minimum at $x = 22\%$. Although the intensity should vanish completely, the dark line is considerably brighter than the frame whose color is equal to that of the vacuum region at the top of the image. This additional background contribution is due to (amongst other factors) inelastically scattered electrons that produce images that are slightly shifted (or rotated), because the focal length of the objective lens depends on the electron energy. An additional effect arises from the fact that the propagation direction of an inelastically scattered electron may be altered during the inelastic scattering process. Subsequently, the electron wave (which has been changed in energy and direction) is diffracted again in the crystal. Owing to the change in direction of the electron wave, the corresponding beams are separated in reciprocal space. Spherical aberration and the defocus (which is different from the defocus for elastically scattered electrons, as mentioned above) lead to a delocalization in the image plane. The spatial information transferred by, for example, the (002) beam of an inelastically scattered electron is shifted in the image plane with respect to the (002) beam of the elastically scattered electrons. The result is a blurring of the DF image. This effect is decreased if an interference pattern of, for example, the (002) and (000) beams is used. For inelastically scattered electrons, the interference-induced fringe patterns are statistically shifted by the effects described above. The superposition of the fringe patterns strongly decreases (or even wipes out) the fringe contrast, producing an image with a nearly homogenous intensity. The inelastically scattered electrons thus contribute to $J(\mathbf{g}_{000}, \mathbf{r})$, whereas $J(\mathbf{g}_{002}, \mathbf{r})$ is less affected. Note that these considerations apply analogously for all other sources of image shifts, such as spatial and temporal incoherence, which thus also contribute to the blurring of a DF image.

5.1.2.3 Conventional HRTEM We now turn to the question of whether conventional HRTEM carried out in an exact ZA orientation of the specimen could be useful for measuring the local amplitude of the (002) reflection. The complications associated with the exact ZA orientation were pointed out in Sect. 2.3. Figure 2.9a displays a complex dependence of $|F(\mathbf{g}_{002})|$ on the specimen thickness at the exact ZA orientation, which complicates the measurement if thickness fluctuations are present.

In addition, the complex interaction of many excited beams results in a high sensitivity to small variations of the specimen orientation, which can occur because of strain fields or local specimen heating by the incident electron beam.

Another important point is that the composition evaluation is simplest, and thus most accurate, if $J(\mathbf{g}_{002}, \mathbf{r}) \propto |F(\mathbf{g}_{002}, \mathbf{r})|$. This sets restrictions on the number of beams contributing to $J(\mathbf{g}_{002}, \mathbf{r})$, which are usually not satisfied for the many-beam conditions conventionally applied in HRTEM.

5.1.2.4 Thickness Measurement We to note here that a knowledge of the specimen thickness in the electron beam direction may not be necessary. We shall see in Sect. 5.3.1 that the CELFA procedure requires one reference point with a known composition to compute the factor of proportionality between the intensity values supplied by the CCD camera on the one hand and the Bloch wave computations on the other hand. The reference point is most suitably chosen in the substrate or buffer layer. For example, to evaluate an $In_x Ga_{1-x} As/GaAs(001)$ quantum well, we would choose the reference point to be in the GaAs, where $x = 0$. All measured values of $F(\mathbf{g}_{002}, \mathbf{r})$

would be normalized with respect to the modulus $a(\mathbf{g}_{002}, x = 0)$ of the (002) beam measured in the GaAs substrate. An exact knowledge of the specimen thickness is not necessary if $a(\mathbf{g}_{002}, x)/a(\mathbf{g}_{002}, x = 0)$ shows a sufficiently weak dependence on the specimen thickness. We shall see that this condition is well fulfilled for $In_x Ga_{1-x} As$ and $Cd_x Zn_{1-x} Se$. It will be shown later that an estimation of the specimen thickness requires a three-beam condition and the exposure of a defocus series, whereas a two-beam condition and the exposure of only one image are sufficient if a knowledge of the specimen thickness is not essential. Note that the local specimen thickness can also be obtained by electron holography.

5.1.2.5 Suggested Imaging Conditions and Procedure for the Application of the CELFA Technique
The CELFA procedure requires the following steps:

1. The specimen is tilted 2–4° with respect to an exact $\langle 100 \rangle$ ZA orientation, with either the (004) or the (002) beam strongly excited.
2. The (002) beam is centered on the optical axis to minimize the impact of objective lens aberrations.
3. The (000) and (002) beams are selected with the objective lens aperture if a knowledge of the specimen thickness is not necessary. Otherwise, the three beams (000), (002) and (004) are chosen. It will be shown later that the two-beam and three-beam conditions fulfill the condition $J(\mathbf{g}_{002}, \mathbf{r}) \propto |F(\mathbf{g}_{002}, \mathbf{r})|$.
4. The images are recorded with an on-line CCD camera. One single image is recorded if the two-beam condition is chosen. Otherwise, a defocus series consisting of about 10 to 20 images is taken. The defocus step size should be less than about 10 nm for an acceleration voltage of 200 keV.
5. If required, a noise reduction is performed as described in Sect. 4.1.1.
6. Local values $A_S(\mathbf{g}_{002}, \mathbf{r})$ are measured from the experimental image.
7. The local composition is obtained by comparing the experimentally measured values $A_S(\mathbf{g}_{002}, \mathbf{r})$ with values $A_S^{\text{Bloch}}(\mathbf{g}_{002}, \mathbf{r})$ based on Bloch wave calculations and nonlinear imaging theory.

5.1.3 Fringe Images

In this chapter, we illustrate the procedure with an example different from that used in Chap. 4. Here, we use a buried SK layer grown on a GaAs (001) substrate and capped with 10 nm GaAs. Figure 5.2a depicts a fringe image, which is the first image of a defocus series of ten images. The defocus step size was adjusted to 9 nm. A noise reduction was performed as described in Sect. 4.1.1. Additionally, an inverse Fourier transformation of the noise-reduced Fourier transform was performed by placing circular apertures around the (002) and (004) reflections. The radii of the circular apertures were chosen in such a manner that the circles overlapped. Inside the GaAs buffer and cap

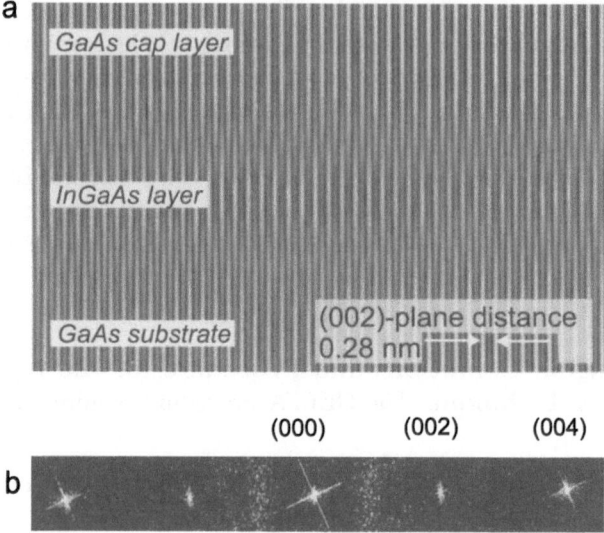

Fig. 5.2. (a) HRTEM fringe image of an $In_x Ga_{1-x} As$ layer buried in GaAs obtained with the imaging condition described in Sect. 5.1.2. The spacing of the *bright lines* corresponds to the (002) plane spacing of 0.28 nm. (b) Central part of the corresponding diffractogram, showing the (000), $\pm$(002) and $\pm$(004) reflections

layer, Fig. 5.2a reveals a contrast pattern consisting of alternating bright, dark, less bright and again dark fringes. The bright fringes correspond to the (002) lattice planes. Their spacing is about 0.28 nm. With increasing In concentration (from the bottom to the middle of Fig. 5.2a), the intensity of the bright fringes decreases, whereas the brightness of the darker fringes increases. Their intensities become equal at an In concentration of about 22%, where the (002) lattice fringes vanish owing to the vanishing (002) structure factor, and only the (004) fringes are visible. From Fig. 5.2a we can see, that the imaging condition used is well suited for the compositional analysis that we shall describe. Figure 5.2b shows the central part of the diffractogram, containing the (000), $\pm$(002) and $\pm$(004) reflections. Note that only the (000), (002) and (004) beams contribute to the image formation. The next section provides an analytical relationship between the reflections in Fig. 5.2b and the contributing beams.

5.2 Theoretical Considerations

5.2.1 Calculation of Reflection Amplitudes

We start with the consideration of the (000), (002) and (004) beams that contribute to the image formation. All other ZOLZ beams are excluded by the

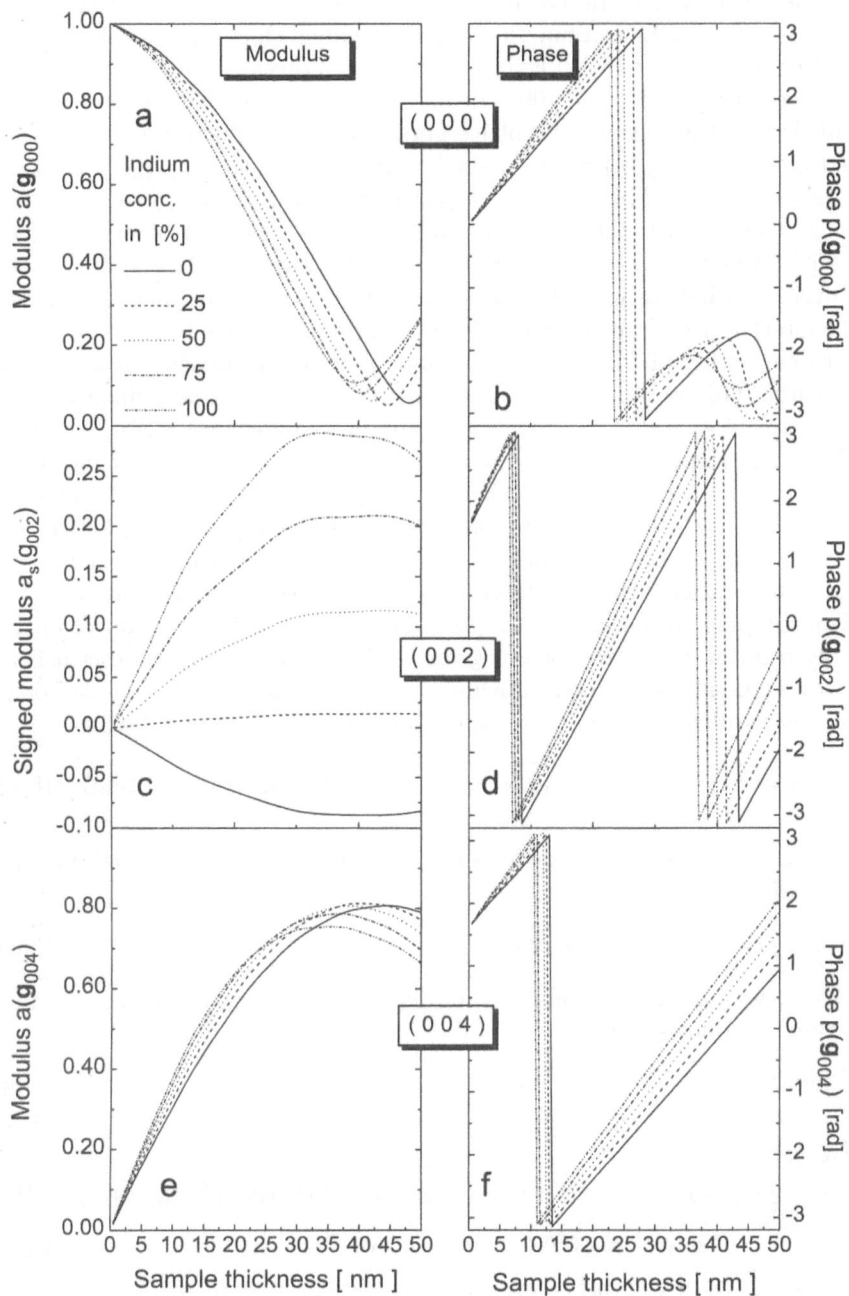

Fig. 5.3. Results of Bloch wave computations performed with the EMS program package for $In_xGa_{1-x}As$ with different indium concentrations. (a) Modulus of the (000) beam, (c) signed modulus of the (002) beam and (e) modulus of the (004) beam, plotted versus the sample thickness; (b), (d) and (f) show the corresponding phases

objective aperture. Figure 5.3 shows the amplitudes and phases of the three beams in $In_xGa_{1-x}As$ as a function of the specimen thickness and the In concentration x, calculated with EMS [2]. Figure 5.3 reveals that the amplitudes and phases of the (000) and (004) beams show a rather weak dependence on x. This is also the case for the phase $p(\mathbf{g}_{002})$ of the (002) beam, whereas its signed modulus $a_S(\mathbf{g}_{002})$ varies strongly with x. Note that $a_S(\mathbf{g}_{002})$ changes its sign as x varies. This is due to a phase shift of π, which, for clarity, has been associated with the signed modulus $a_S(\mathbf{g}_{002})$ instead of $p(\mathbf{g}_{002})$. For $x = 25\%$, the curve $a_S(\mathbf{g}_{002}, t)$ (t denotes the sample thickness) is close to zero over the whole sample thickness range. In this case, the image is formed by the interference of the (000) and (004) beams only, which leads to an image pattern consisting of fringes with the same maximum intensity at a spacing of 0.14 nm. This is the case for all sample thicknesses t and defocus values Δf. Figures (5.4) and (5.5) show the moduli and phases for $Cd_xZn_{1-x}Se$ and $Al_xGa_{1-x}As$, respectively. In $Cd_xZn_{1-x}Se$, the signed modulus of the (002) beam vanishes at a Cd concentration of approximately $x = 40\%$. In contrast, $Al_xGa_{1-x}As$ represents a material where the sign of the (002) beam remains positive over the whole range of x. In the following, the nonlinear image formation is treated in detail for the conditions described above.

The complex amplitude of the reflection $\mathbf{g}_p$ in the diffractogram is, in the case of an untilted electron beam, given by (3.36). However, as described in Sect. 5.1.2, the incident electron beam was tilted in such a way that the (002) beam was parallel to the optic axis. In this case, (3.36) is modified to:

$$J(\mathbf{g}_p) = \sum_m T(\mathbf{g}_p + \mathbf{g}_m - \mathbf{g}_{002}, \mathbf{g}_m - \mathbf{g}_{002}; \Delta f) F(\mathbf{g}_p + \mathbf{g}_m) F^*(\mathbf{g}_m) . \quad (5.7)$$

As mentioned above, we consider only the (000), (002) and (004) beams, which leads us to

$$F(\mathbf{g}_{00l}) = 0 \quad \text{for} \quad l \notin \{0, 2, 4\} . \tag{5.8}$$

Therefore, we obtain the complex Fourier coefficients of the three relevant reflections as follows:

$$J(\mathbf{g}_{000}) = \sum_{l=0,2,4} F(\mathbf{g}_{00l}) F^*(\mathbf{g}_{00l}) , \tag{5.9}$$

$$J(\mathbf{g}_{002}) =$$
$$T(0, -\mathbf{g}_{002}; \Delta f) F(\mathbf{g}_{002}) F^*(0) + T(\mathbf{g}_{002}, 0; \Delta f) F(\mathbf{g}_{004}) F^*(\mathbf{g}_{002}) , \tag{5.10}$$

$$J(\mathbf{g}_{004}) = T(\mathbf{g}_{002}, -\mathbf{g}_{002}; \Delta f) F(\mathbf{g}_{004}) F^*(0) . \tag{5.11}$$

According to (3.38) and (3.39) the transmission cross-coefficients in (5.10) are connected by the relations

$$T(0, -\mathbf{g}_{002}; \Delta f) \stackrel{(3.38)}{=} T^*(-\mathbf{g}_{002}, 0; \Delta f)$$
$$\stackrel{(3.39)}{=} T^*(\mathbf{g}_{002}, 0; \Delta f) . \tag{5.12}$$

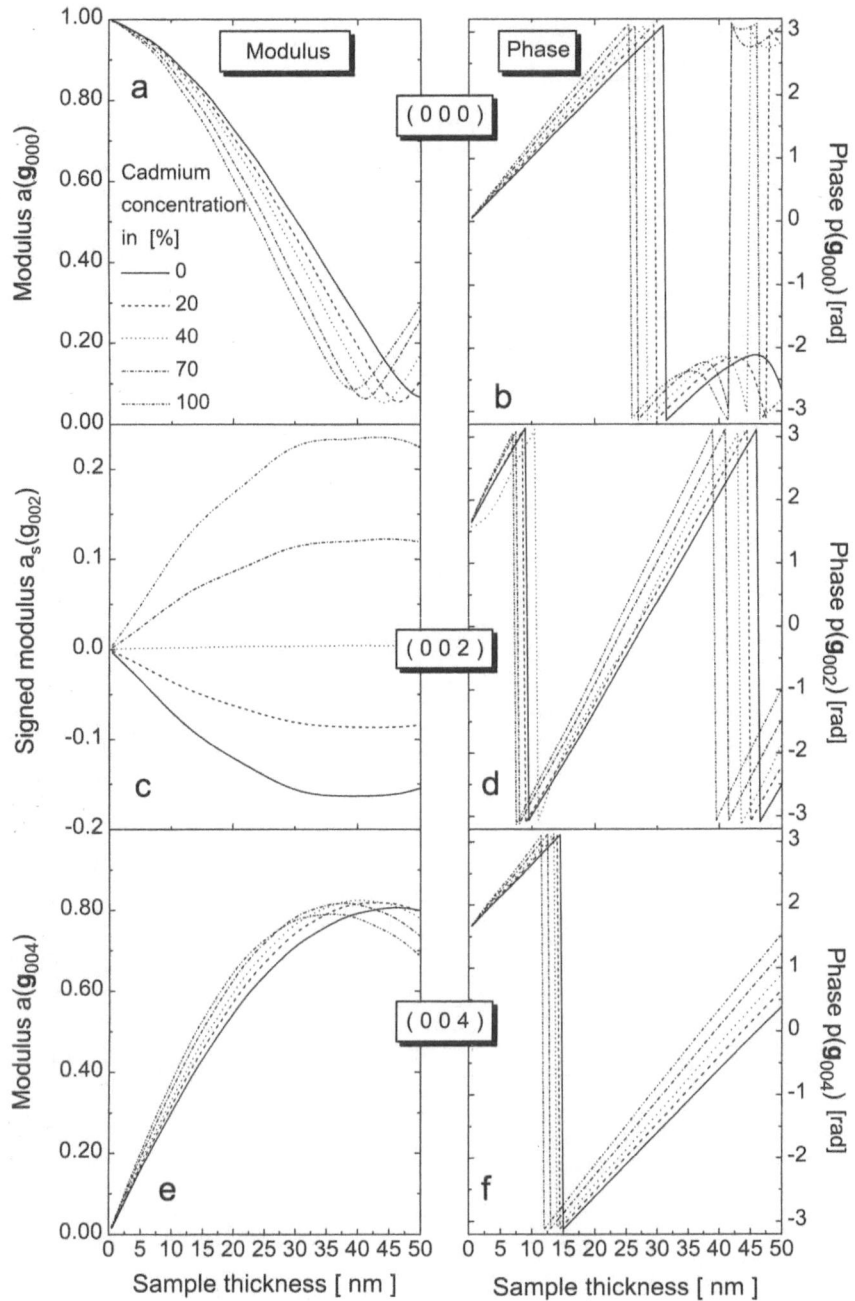

Fig. 5.4. Results of Bloch wave computations performed with the EMS program package for $Cd_xZn_{1-x}Se$ with different indium concentrations. (**a**) Modulus of the (000) beam, (**c**) signed modulus of the (002) beam and (**e**) modulus of the (004) beam, plotted versus the sample thickness; (**b**), (**d**) and (**f**) show the corresponding phases

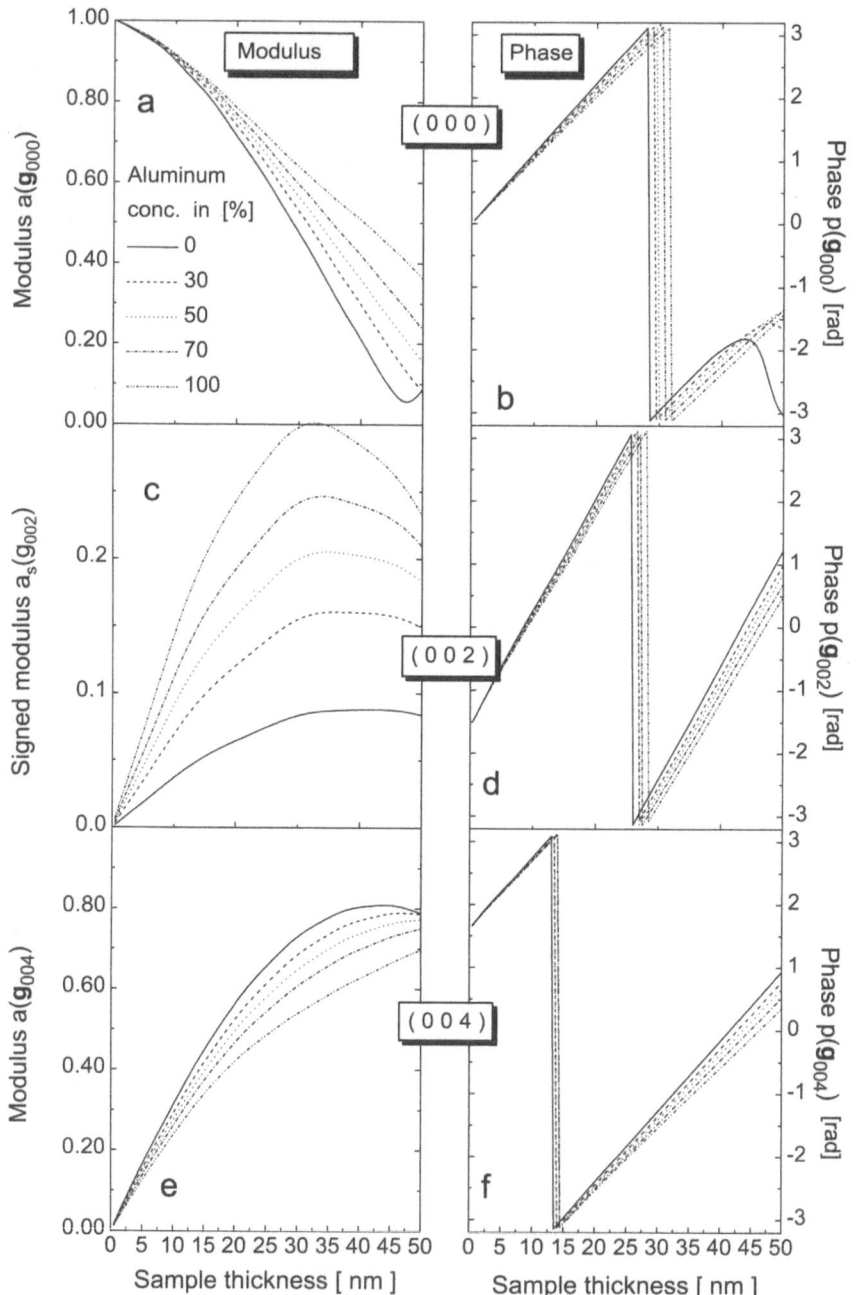

Fig. 5.5. Results of Bloch wave computations performed with the EMS program package for $Al_xGa_{1-x}As$ with different indium concentrations. (**a**) Modulus of the (000) beam, (**c**) signed modulus of the (002) beam and (**e**) modulus of the (004) beam, plotted versus the sample thickness; (**b**), (**d**) and (**f**) show the corresponding phases

The quantity $T(\mathbf{g}_{002}, -\mathbf{g}_{002}; \Delta f)$ in (5.11) is real because

$$T(\mathbf{g}_{002}, -\mathbf{g}_{002}; \Delta f) \overset{(3.38)}{=} T^{\star}(-\mathbf{g}_{002}, \mathbf{g}_{002}; \Delta f)$$

$$\overset{(3.39)}{=} T^{\star}(\mathbf{g}_{002}, -\mathbf{g}_{002}; \Delta f) . \tag{5.13}$$

Therefore, we may use the abbreviations

$$T(\mathbf{g}_{002}, 0; \Delta f) = T_{002} \exp\{i\chi\} ,$$
$$T(0, -\mathbf{g}_{002}; \Delta f) = T_{002} \exp\{-i\chi\} ,$$
$$T(\mathbf{g}_{002}, -\mathbf{g}_{002}; \Delta f) = T_{004} , \tag{5.14}$$

where T_{002} and T_{004} are real numbers and χ is the phase shift introduced by the objective lens defocus and the spherical aberration, given by (3.10). Inserting the abbreviations defined in (5.14) into (5.10) and (5.11) yields

$$J(\mathbf{g}_{002}) = T_{002} \left[e^{-i\chi} F(\mathbf{g}_{002}) F^{\star}(0) + e^{i\chi} F(\mathbf{g}_{004}) F^{\star}(\mathbf{g}_{002}) , \right] \tag{5.15}$$

$$J(\mathbf{g}_{004}) = T_{004} F(\mathbf{g}_{004}) F^{\star}(0) . \tag{5.16}$$

Introducing the (signed) moduli and phases of the beams, we finally obtain the moduli $A(\mathbf{g}_{00l}) = |J(\mathbf{g}_{00l})|$ of the reflections as follows:

$$\boxed{\begin{aligned} &A_{\mathrm{S}}(\mathbf{g}_{002}) \\ &= T_{002} a_{\mathrm{S}}(\mathbf{g}_{002}) \sqrt{a(\mathbf{g}_{000})^2 + a(\mathbf{g}_{004})^2 + 2a(\mathbf{g}_{000}) a(\mathbf{g}_{004}) \cos \varphi} , \\ &A(\mathbf{g}_{004}) = T_{004} a(\mathbf{g}_{004}) a(\mathbf{g}_{000}) , \end{aligned}}$$

$$\tag{5.17}$$

where φ is given by

$$\varphi = -2\chi + 2p(\mathbf{g}_{002}) - p(\mathbf{g}_{000}) - p(\mathbf{g}_{004}) . \tag{5.18}$$

Equations (5.18), (5.17), (3.10) and (3.4) show that a linear change of the defocus results in an oscillation of $A_{\mathrm{S}}(\mathbf{g}_{002})$. The defocus change $\delta(\Delta f)$ corresponding to a full oscillation of $A_{\mathrm{S}}(\mathbf{g}_{002})$ is given by

$$\delta(\Delta f) = \frac{1}{\lambda \mathbf{g}_{002}^2} . \tag{5.19}$$

Note that the results for a two-beam condition where only the (000) and (002) beams are selected by the objective lens aperture can be obtained simply from (5.17) by putting $a(\mathbf{g}_{004}) \equiv 0$.

5.2.2 Determination of Sample Thickness and Phase φ

The determination of the specimen thickness presented here is based on the acquisition of a defocus series in an area of known composition. Equation

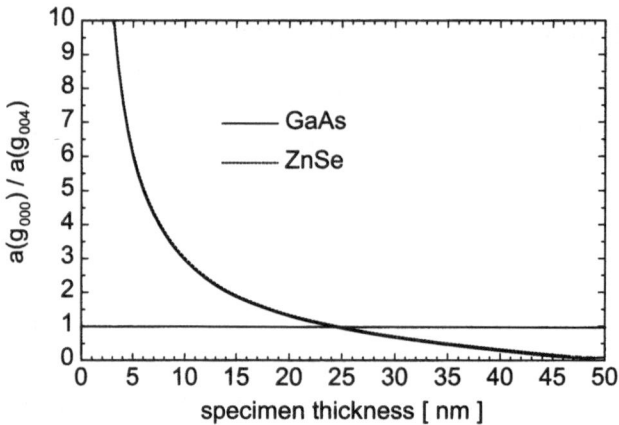

Fig. 5.6. Ratio of the moduli of the (000) and (004) beams plotted versus the sample thickness. This graph can be used to determine the specimen thickness. The excitation conditions were chosen as described in Sect. 5.1.2

(5.17) indicates an oscillation of $A_S(\mathbf{g}_{002})$ as a function of φ, which is a linear function of the defocus. The basic idea is to measure the maximum and minimum values A_{max} and A_{min} of the oscillating quantity $A_S(\mathbf{g}_{002})$, and these values can then be used to obtain the specimen thickness, as will be shown in the following. According to (5.17), the oscillation of $A_S(\mathbf{g}_{002})$ has its extrema at $\cos\varphi = \pm 1$. The extreme amplitudes of the (002) reflection are thus given by

$$A_{extrema} = T_{002} a(\mathbf{g}_{002}) \left| a(\mathbf{g}_{000}) \pm a(\mathbf{g}_{004}) \right| . \tag{5.20}$$

We now consider

$$\frac{A_{max} + A_{min}}{A_{max} - A_{min}} = \begin{cases} a(\mathbf{g}_{000})/a(\mathbf{g}_{004}) & \text{for } a(\mathbf{g}_{000}) > a(\mathbf{g}_{004}) \\ a(\mathbf{g}_{004})/a(\mathbf{g}_{000}) & \text{otherwise} \end{cases} . \tag{5.21}$$

Figure 5.6 shows that $a(\mathbf{g}_{000})/a(\mathbf{g}_{004})$ is a continuous function of the specimen thickness, so that a measurement of $(A_{max} + A_{min})/(A_{max} - A_{min})$ can be used for determination of the thickness. This procedure is unambiguous if the specimen thickness is below a certain value, for example 25 nm for GaAs and ZnSe. This limit can be shifted by varying the imaging conditions.

Figure 5.7 gives an example of a defocus series consisting of ten images (nine of them are shown), where the defocus step size of the microscope was adjusted to 9 nm. Each image is a small part of an image of size 1024×1024 pixels^2. An area of known composition in the GaAs buffer layer was used for the thickness determination. Figure 5.8 shows the amplitudes $A_S(\mathbf{g}_{002})$ plotted versus the image number n (running from $n = 1$ to 10, 1 corresponding to the largest underfocus). Figure 5.8 was obtained in the following way. From each of the ten images, a region of the same size and position was Fourier-transformed. The ten pixels with the largest amplitudes in a circular area

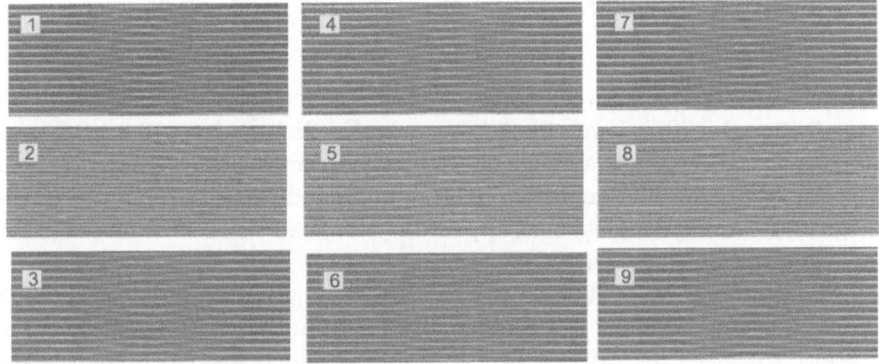

Fig. 5.7. Small parts of HRTEM images in a defocus series of ten images (nine of them are shown), each showing the $In_xGa_{1-x}As$ interlayer as well as the GaAs cap and buffer layers. A defocus step size of 9 nm was chosen. Images 2, 5 and 8 show a fringe pattern that is dominated by the (004) fringes, corresponding to minima of the amplitude of the (002) reflection shown in Fig. 5.8

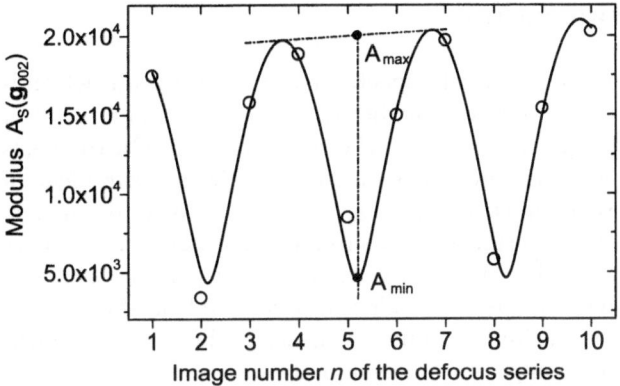

Fig. 5.8. Amplitude of the (002) reflection in the diffractogram plotted versus the image number shown in Fig. 5.7. The values A_{max} and A_{min} were used to derive the specimen thickness

around the (002) reflection were summed for each image. The data points were fitted by a curve given by

$$A_S(\mathbf{g}_{002}) = (1 + n\alpha)\sqrt{B + C\cos\left(D(n - E)\right)}, \quad n = 1, \ldots, 10 , \qquad (5.22)$$

where α, B, C, D and E are the fit parameters and n is the image number. The factor α in (5.22) takes into account the (weak) defocus dependence of T_{002} in (5.17), which contains the source-size-dependent envelope function $E_S(\mathbf{g}_{002}, 0; \Delta f)$ given by (3.26). The values obtained for the fit parameters were

$$\alpha = 0.012 \,, \quad B = 1.88 \times 10^8 \,, \quad C = 1.7 \times 10^8 \,, \quad D = 2.06 \pm 0.04 \,,$$
$$E = 0.6 \,. \tag{5.23}$$

The defocus step size between adjacent images in the defocus series calculated using 5.19 and 5.22, is given by $\Delta f_{\text{step size}} = D/(2\pi\lambda \mathbf{g}_{002}^2) = (10.3 \pm 0.2)$ nm, which is in rather good agreement with the experimentally chosen defocus step size of 9 nm. The angles φ_n (the index n indicates the number in the image of the defocus series) that appear in (5.17) can be obtained from (5.22) and are given by $\varphi_n = D(n - E)$.

From Fig. 5.8, we deduce a value of $(A_{\max} + A_{\min})/(A_{\max} - A_{\min}) = 1.6$. The sample thickness can be obtained directly from the thickness dependence of $a(\mathbf{g}_{000})/a(\mathbf{g}_{004})$ for GaAs shown in Fig. 5.6, which was calculated by the Bloch wave method. Figure 5.6 also contains data for ZnSe. The calculations were performed with an absorption coeffiecient of 0.04. However, the curves do not depend on the absorption, to a good approximation. In the present case, we find a sample thickness of 16 nm.

5.3 Practical Considerations

Here we focus on some more practical aspects of the composition determination. First we consider the relationship between the local amplitudes of reflections and the "shape" of reflections in the diffractogram. On the basis of these considerations, we then present a method for the measurement of the local amplitudes of reflections, which is applied to the example image. Subsequently, we point out alternative measurement methods and discuss their advantages and drawbacks. We then concentrate on possible errors in the composition determination, where we address objective lens aberrations, noise reduction and, finally, the impact of uncertainties in the specimen thickness on the accuracy of the composition determination.

5.3.1 Measurement Procedure

On the basis of the theoretical considerations in Sect. 5.2, we formulate a more detailed procedure for composition determination:

1. The central point of the procedure is the measurement of *normalized signed local moduli*, given by

$$A_{\text{NS}}(\mathbf{g}_{002}, \mathbf{r}) = A_{\text{S}}(\mathbf{g}_{002}, \mathbf{r})/A_{\text{S}}^{\text{Ref}}(\mathbf{g}_{002}, \mathbf{r}_0) \,. \tag{5.24}$$

A detailed description of the measurement procedure will be given later in this section. The normalization is performed relative to a reference value $A_{\text{S}}^{\text{Ref}}(\mathbf{g}_{002}, \mathbf{r}_0)$, which is obtained from a region of the image around $\mathbf{r}_0$ where the composition x is known. Most conveniently, the reference value is obtained from a region in the substrate where $x = 0$.

2. Bloch wave calculations are carried out using parameters corresponding to the specimen orientation and acceleration voltage of the electron beam used in the experiment, yielding calculated beam amplitudes $F(\mathbf{g}_{00l}, x, t)$, $l = 0, 2, 4$, where the concentration x is varied between 0 and 1 with a step size of 0.01, and thickness values t are taken over the interval 0.5 to 70 nm with a step size of 0.5 nm. Signed reflection moduli $A_S(\mathbf{g}_{002}, x, t)$ are computed from the result using (5.17), and normalized signed moduli are obtained according to

$$A_{NS}^{Bloch}(x, t) = A_S(\mathbf{g}_{002}, x, t)/A_S(\mathbf{g}_{002}, x = 0, t) . \qquad (5.25)$$

3. Each measured value $A_{NS}(\mathbf{g}_{002}, \mathbf{r})$ is compared with the table of simulated values $A_{NS}^{Bloch}(x, t)$. A specimen thickness t is used that is either estimated roughly or obtained as described in Sect. 5.2.2. The value of x that yields the best agreement of $A_{NS}(\mathbf{g}_{002}, \mathbf{r})$ with $A_{NS}^{Bloch}(x)$ is the composition, determined by this method.

We now need a procedure for the measurement of the local reflection amplitude $J(\mathbf{g}_{002}, \mathbf{r})$, from which we obtain $A_{NS}(\mathbf{g}_{002}, \mathbf{r})$ according to (5.24) and (5.5), as required for the first step of the above procedure. In the following we explain the most straightforward approach, based upon Fourier filtering. Alternative methods will be discussed in Sect. 5.3.3.

5.3.1.1 The Shape Function The method presented here is based on the processing of the image diffractogram. As a first step, it is instructive to consider the relationship between a local variation of the reflection amplitudes within the image and the "shape" of the reflections in the diffractogram. The deviation of the local amplitudes $J(\mathbf{g}_p, \mathbf{r})$ from their reference values $J(\mathbf{g}_p, \mathbf{r}_0)$ at a reference position $\mathbf{r}_0$ can be described by a function $S_p(\mathbf{r})$ as follows:

$$J(\mathbf{g}_p, \mathbf{r}) = J(\mathbf{g}_p, \mathbf{r}_0)S_p(\mathbf{r}) . \qquad (5.26)$$

Note that $S_p(\mathbf{r})$ contains effects of local variations of both the modulus and the phase of $J(\mathbf{g}_p, \mathbf{r})$. A local gradient of the phase may arise from a varying lattice parameter, for example. Obviously, $S_p(\mathbf{r})$ carries all the spatial information we are interested in. It can be regarded as a "shape function" holding information about the geometry and local composition of the nanostructures contained in the image. We now consider the contribution of $J(\mathbf{g}_p, \mathbf{r})$ to the diffractogram. According to (5.6), the contribution of $J(\mathbf{g}_p, \mathbf{r})$ to the image intensity is given by

$$\tilde{I}_{\mathbf{g}_p}(\mathbf{r}) = J(\mathbf{g}_p, \mathbf{r}_0)S_p(\mathbf{r})\exp\{-2\pi i \mathbf{g}_p \cdot \mathbf{r}\} , \qquad (5.27)$$

which yields

$$\tilde{I}_{\mathbf{g}_p}(\mathbf{k}) = \int J(\mathbf{g}_p, \mathbf{r}_0)S_p(\mathbf{r})\exp\{2\pi i(\mathbf{k} - \mathbf{g}_p) \cdot \mathbf{r}\}\, d^2\mathbf{r} . \qquad (5.28)$$

For a spatially homogeneous specimen, $S_p(\mathbf{r}) = 1$, and the contribution of the reflection $\mathbf{g}_p$ to the diffractogram would result in $\widetilde{I}_{\mathbf{g}_p}(\mathbf{k}) = J(\mathbf{g}_p, \mathbf{r}_0)\delta(\mathbf{k} - \mathbf{g}_p)$ from (5.28); this result is identical to (3.15) if we describe the whole diffractogram by

$$\widetilde{I}(\mathbf{k}) = \sum_p \widetilde{I}_{\mathbf{g}_p}(\mathbf{k}) \ . \tag{5.29}$$

In a nonhomogeneous specimen, $\widetilde{I}_{\mathbf{g}_p}(\mathbf{k})$ is not a δ-"function" anymore. Rewriting (5.28) in the form

$$\frac{\widetilde{I}_{\mathbf{g}_p}(\mathbf{k})}{J(\mathbf{g}_p, \mathbf{r}_0)} = \mathcal{F}\left[S_p(\mathbf{r}) \exp\{-2\pi i \mathbf{g}_p \cdot \mathbf{r}\}\right]$$

$$= [\mathcal{F}S_p(\mathbf{r})] \otimes [\mathcal{F}\exp\{-2\pi i \mathbf{g}_p \cdot \mathbf{r}\}]$$

$$= [\mathcal{F}S_p(\mathbf{r})] \otimes \delta(\mathbf{k} - \mathbf{g}_p) \ , \tag{5.30}$$

we recognize that the δ-"function" that occurs for a homogenous specimen is replaced by the convolution $[\mathcal{F}S_p(\mathbf{r})] \otimes \delta(\mathbf{k} - \mathbf{g}_p)$. Depending on factors such as the abruptness of chemical transitions and the presence of specimen edges in the image, $\widetilde{I}_{\mathbf{g}_p}(\mathbf{k})$ may have nonvanishing components at positions $\mathbf{k}$ of the Fourier space that are significantly separated from $\mathbf{k} = \mathbf{g}_p$.

5.3.1.2 Measurement of $J(\mathbf{g}_p, \mathbf{r})$ The Fourier filtering method for the measurement of $J(\mathbf{g}_{002}, \mathbf{r})$ presented in this section is based on the presupposition that all (significant) components of $\widetilde{I}_{\mathbf{g}_p}(\mathbf{k})$ are found in a circular area $|\mathbf{g}_p - \mathbf{k}| < R$ where the radius R is smaller than half the distance to any of the neighboring reflections. In this case, $\widetilde{I}_{\mathbf{g}_p}(\mathbf{k})$ can be selected from the diffractogram $\widetilde{I}(\mathbf{k})$ by deleting all diffractogram information at positions $|\mathbf{g}_p - \mathbf{k}| \geq R$, as depicted in Fig. 5.9b. Figure 5.9c shows the filtered diffractogram. Next, we have to compute $J(\mathbf{g}_p, \mathbf{r})$. Solving (5.30) with respect to $S_p(\mathbf{r})J(\mathbf{g}_p, \mathbf{r}_0)$, we find

$$J(\mathbf{g}_p, \mathbf{r}_0)S_p(\mathbf{r}) = \exp\{2\pi i \mathbf{g}_p \cdot \mathbf{r}\} \left[\mathcal{F}^{-1}\widetilde{I}_{\mathbf{g}_p}(\mathbf{k})\right]$$

$$= \mathcal{F}^{-1}\left[\mathcal{F}\exp\{2\pi i \mathbf{g}_p \cdot \mathbf{r}\}\right] \cdot \mathcal{F}^{-1}\widetilde{I}_{\mathbf{g}_p}(\mathbf{k})$$

$$= \mathcal{F}^{-1}\left[\delta(\mathbf{k} + \mathbf{g}_p) \otimes \widetilde{I}_{\mathbf{g}_p}(\mathbf{k})\right] \ . \tag{5.31}$$

Equation (5.31) can be used to derive a practical method for the computation of $J(\mathbf{g}_p, \mathbf{r})$. The convolution of $\widetilde{I}_{\mathbf{g}_p}(\mathbf{k})$ with $\delta(\mathbf{k} + \mathbf{g}_p)$ corresponds to a shift of $\widetilde{I}_{\mathbf{g}_p}(\mathbf{k})$ by $-\mathbf{g}_p$ in Fourier space. This "centering" of the reflection $\mathbf{g}_p$ is visualized in Fig. 5.9d. The subsequent step, also indicated in (5.31), is an inverse Fourier transform. The result is the complex-valued local amplitude $J(\mathbf{g}_p, \mathbf{r})$ of the reflection $\mathbf{g}_p$, from which we obtain for the special case $\mathbf{g} = \mathbf{g}_{002}$ the normalized signed local modulus $A_{NS}(\mathbf{g}_{002}, \mathbf{r})$. Note that the sign of $A_{NS}(\mathbf{g}_{002}, \mathbf{r})$ is evaluated by processing the phase of $J(\mathbf{g}_{002}, \mathbf{r})$, as will be

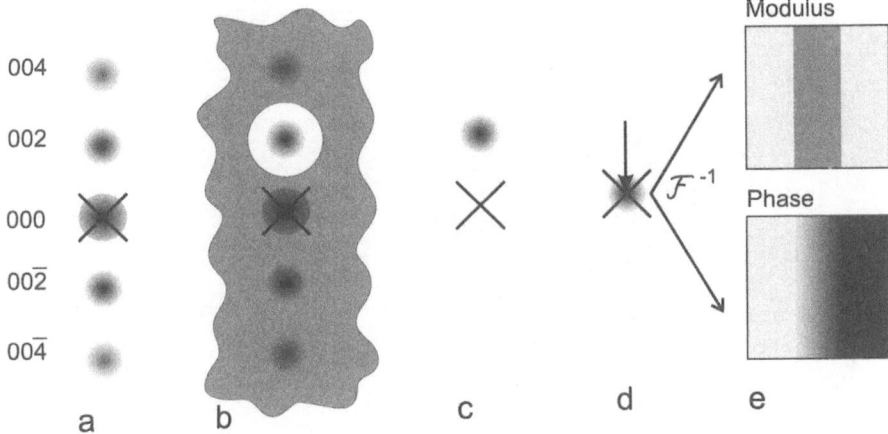

Fig. 5.9. Schematic drawing explainsing the Fourier filtering method used to measure local reflection amplitudes $J(\mathbf{g}_p, \mathbf{r})$. (a) The diffractogram obtained in the three-beam condition used for the CELFA analysis, which contains the reflections (000), $\pm$(002) and $\pm$(004). The *cross* marks the center of Fourier space $\mathbf{k} = 0$. (b) The filtering of the diffractogram, where only the information in a circular area around the (002) reflection is kept, and all information outside the circle is deleted. (c) The filtered diffractogram. (d) Result obtained after the centering of the (002) reflection. (e) The local modulus and the local phase of the (002) reflection are computed by an inverse Fourier transform

shown in the next section, which presents an example of an evaluation to visualize the analysis procedure.

5.3.2 Example of an Evaluation

Figure 5.10 shows an image of an $In_xGa_{1-x}As/GaAs(001)$ layer obtained under the imaging conditions suggested in Sect. 5.1.2. It is the first image of the defocus series displayed in Fig. 5.7. One can clearly recognize the $In_xGa_{1-x}As$ layer by the half-period shift of the bright (002) lattice fringes. This shift occurs in regions where the In concentration is larger than x_0 because the phase of the (002) beam jumps by π if the In concentration exceeds $x_0 \approx 22\%$.

5.3.2.1 Local Modulus and Phase The application of the procedure visualized in Fig. 5.9 yields the local modulus $A(\mathbf{g}_{002}, \mathbf{r})$ and phase $P(\mathbf{g}_{002}, \mathbf{r})$ of the (002) reflection, which are displayed in Figs. 5.11a,b. The most noticeable feature in Fig. 5.11a is the black line that surrounds the $In_xGa_{1-x}As$ region. This line is due to the vanishing local modulus of the (002) reflection that occurs for an In concentration of $x_0 \approx 22\%$. Inside the region bordered by the black line, the In concentration is most probably x greater than x_0,

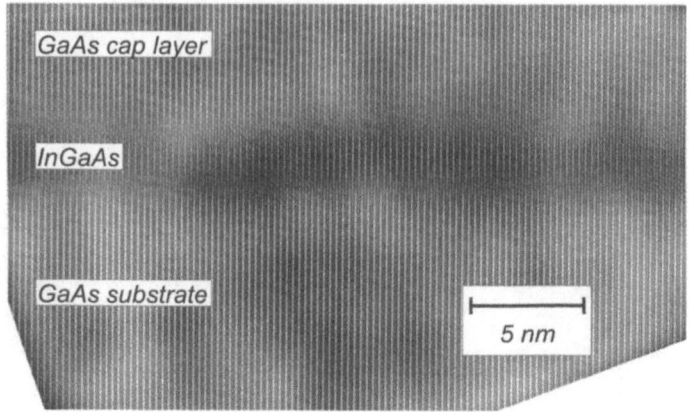

Fig. 5.10. Noise-reduced three-beam HRTEM image of an $In_xGa_{1-x}As/GaAs(001)$ heterostructure, used as an example to illustrate the evaluation procedure. In addition to the dominant (002) and (004) lattice fringes, the image contains a faint dot pattern due to weakly excited {022} reflections, which were also allowed to pass through the objective lens aperture. These {022} do not have any influence on the composition evaluation

whereas it is smaller than x_0 outside. Figure 5.11b shows that this assumption is indeed valid. This figure contains a dark region that seems to coincide with the area surrounded by the black line in Fig. 5.11a. On closer inspection we find a phase difference of π between the dark and bright regions in Fig. 5.11b. This phase difference has been used to compute the sign needed for the calculation of the local signed amplitude $A_S(\mathbf{g}_{002}, \mathbf{r})$. Using (5.5), we obtain $A_S(\mathbf{g}_{002}, \mathbf{r}) = +A(\mathbf{g}_{002}, \mathbf{r})$ inside the dark region in Fig. 5.11b and $A_S(\mathbf{g}_{002}, \mathbf{r}) = -A(\mathbf{g}_{002}, \mathbf{r})$ elsewhere.

5.3.2.2 Local Reference Modulus The next step is to calculate the normalized signed moduli $A_{NS}(\mathbf{g}_{002}, \mathbf{r})$ defined in (5.24). This could be done by considering a reference modulus $A_S^{Ref}(\mathbf{g}_{002}, \mathbf{r}_0)$ at a position $\mathbf{r}_0$ in the GaAs substrate. However, Fig. 5.11a shows that the the local modulus is constant neither in the GaAs substrate nor in the cap layer. Consequently, we cannot find a value of $A_S^{Ref}(\mathbf{g}_{002}, \mathbf{r}_0)$ that is generally valid in the GaAs regions of the image. The small but detectable fluctuations that can be seen in the GaAs regions of Fig. 5.11a could be caused by small fluctuations of the specimen thickness and of the local defocus. , for example. Equation (5.17) shows that a locally varying defocus would modulate the phase φ of (5.17), which would lead to a fluctuation of $J(\mathbf{g}_{002}, \mathbf{r})$. As a consequence, a locally defined value $A_S^{Ref}(\mathbf{g}_{002}, \mathbf{r})$ is desirable for normalization. It is obvious that $A_S^{Ref}(\mathbf{g}_{002}, \mathbf{r})$ is required to be identical to $A_S(\mathbf{g}_{002}, \mathbf{r})$ in regions where $x = 0$. In regions where $x > 0$, the reference modulus cannot be computed directly and thus has to be interpolated from the GaAs substrate and cap layer regions. By this

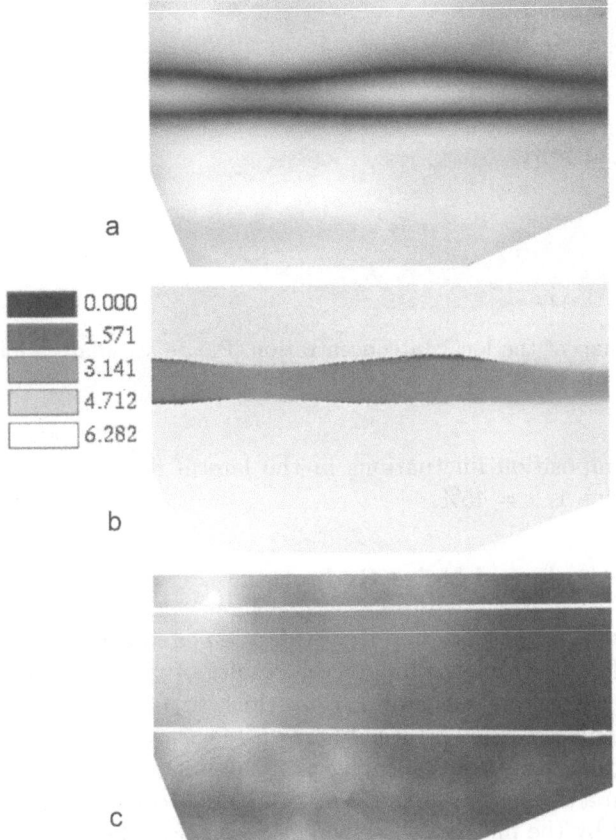

Fig. 5.11. Grayscale-coded maps of (**a**) the local modulus $A(\mathbf{g}_{002}, \mathbf{r})$ and (**b**) the local phase $P(\mathbf{g}_{002}, \mathbf{r})$. (**c**) Grayscale-coded map of the reference modulus $A^{\mathrm{Ref}}(\mathbf{g}_{002}, \mathbf{r})$. The grayscale coding of (a) and (c) has been performed in such a way that black corresponds to the minimum value and white to the maximum

procedure, we have obtained a local reference modulus $A_{\mathrm{S}}^{\mathrm{Ref}}(\mathbf{g}_{002}, \mathbf{r})$, whose absolute value $A^{\mathrm{Ref}}(\mathbf{g}_{002}, \mathbf{r})$ is shown in Fig. 5.11c. The region between the two white lines was chosen for the interpolation. The reference modulus is given by $A_{\mathrm{S}}(\mathbf{g}_{002}, \mathbf{r})$ for positions $\mathbf{r}$ outside this region and was obtained by interpolation in the interior. Finally, $A_{\mathrm{NS}}(\mathbf{g}_{002}, \mathbf{r})$ was computed according to (5.24) from the relation $A_{\mathrm{NS}}(\mathbf{g}_{002}, \mathbf{r}) = A_{\mathrm{S}}(\mathbf{g}_{002}, \mathbf{r})/A_{\mathrm{S}}^{\mathrm{Ref}}(\mathbf{g}_{002}, \mathbf{r})$.

5.3.2.3 Local Composition The final result of the evaluation is the local In concentration. This was obtained by comparison of the experimentally measured values of $A_{\mathrm{NS}}(\mathbf{g}_{002}, \mathbf{r})$ with a table of values of $A_{\mathrm{NS}}^{\mathrm{Bloch}}(x)$ computed by Bloch wave calculations in which the In concentration x was varied from 0 to 1 with a step size of 0.01. The result is shown as a color-coded map of the local In concentration in Fig. 5.12. This map reveals the $\mathrm{In}_x\mathrm{Ga}_{1-x}\mathrm{As}$

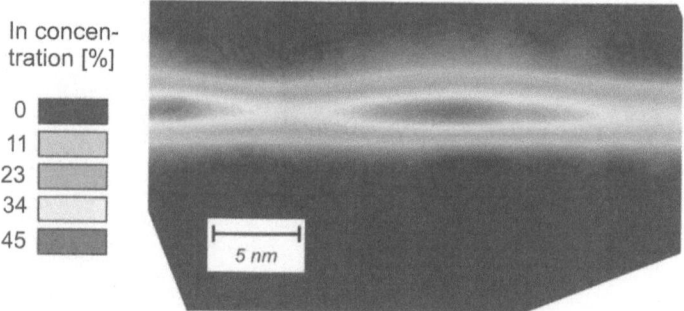

In concentration [%]

0
11
23
34
45

5 nm

Fig. 5.12. Color-coded map of the local In concentration [Please see Plate 9 for color reproduction of this plate]

layer which contains composition fluctuations in the lateral direction. The maximum In concentration is $x = 45\%$.

5.3.3 Alternative Methods and Noise Reduction

We now focus on alternative methods for the evaluation of the local amplitudes of reflections. To allow us to compare the results of the evaluation procedures with known values, an image of a 5 nm thick $In_xGa_{1-x}As$ layer buried in GaAs was simulated (Fig. 5.13a). Figure 5.14a shows profiles of the local amplitude $A(\mathbf{g}_{002}, \mathbf{r})$ of the (002) reflection; the bottom axis corresponds to the vertical direction in Fig. 5.13a. The solid curve corresponds to an evaluation performed by the method described in Sect. 5.3.1. The dashed curve is the true profile. Figure 5.14a clearly demonstrates a drawback of the evaluation procedure of Sect. 5.3.1, because the solid line reveals artificial oscillations in the evaluated profile. Note that the oscillations are strongest in the vicinity of the image borders and close to the transition regions of the $In_xGa_{1-x}As$ interlayer. Two reasons can be found, which are discussed in the following.

5.3.3.1 Drawbacks of the Fourier Filtering Method First, (5.30) shows that the contribution $\widetilde{I}_{\mathbf{g}_p}(\mathbf{k})$ of a "reflection" $\mathbf{g}_p$ to the diffractogram of the image cannot be described by a δ-"function" if the image is not homogenous, but instead is proportional to the convolution of $\delta(\mathbf{k} - \mathbf{g}_p)$ with the Fourier transform of the shape function $S_p(\mathbf{r})$. Selecting a circular area around a reflection in the diffractogram with a mask and deleting all information lying outside may thus partially exclude information about $J(\mathbf{g}_p, \mathbf{r})$. This effect is demonstrated in Fig. 5.15. Figure 5.15a shows a square function $f(r)$ that represents a fictitious shape function. The amplitude of its Fourier transform is depicted in Fig. 5.15b. By selecting the area inside the vertical red lines with a mask, setting the values outside to zero and performing an inverse

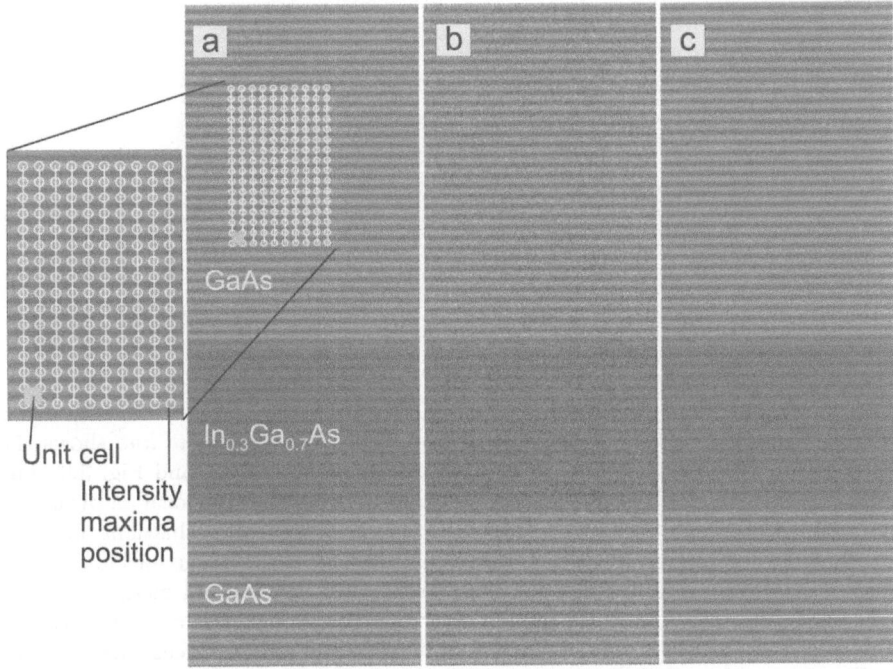

Fig. 5.13. (a) Simulated image of a 5 nm thick $In_xGa_{1-x}As$ layer with an In concentration of 30%, buried in GaAs. An ideal objective lens without aberrations is assumed and only the (000) and (002) beams contribute to the image formation. The transition regions from GaAs to $In_{0.3}Ga_{0.7}As$ and vice versa have a thickness of 2 ML each. The In concentration varies linearly within the transition regions. The upper part of (a) (shown magnified at the *left*) demonstrates the decomposition of the image into image unit cells. The position of intensity maxima (marked with *circles*) are searched for along the *vertical white lines*. Each unit cell is defined by four detected intensity maxima positions lying at its corners. (b) Image (a) with additional Poisson-distributed noise. (c) Image (a) after noise reduction by the Wiener filtering method presented in Sect. 4.1.1

Fourier transform, we obtain the red curve in Fig. 5.15a. Clearly, the filtering in Fourier space causes artificial oscillations.

The second reason for artificial oscillations, which in this case occur predominantly close to the image border, is also related to the filtering in Fourier space. This reason is due to the fact that the HRTEM image contains only a rectangular part $I(\mathbf{r})$ of the real intensity distribution $I_\infty(\mathbf{r})$. The image can be described by

$$I(\mathbf{r}) = I_\infty(\mathbf{r})W(\mathbf{r}) , \tag{5.32}$$

where $W(\mathbf{r})$ is the *window function*, equal to 1 inside the area depicted in the HRTEM image and 0 elsewhere. The corresponding diffractogram is given by

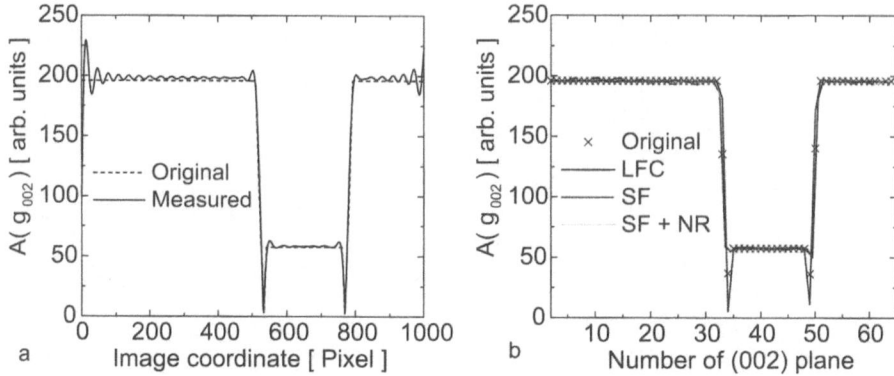

Fig. 5.14. (a) Local amplitude of the (002) reflection $A(g_{002}, r)$ plotted versus the pixel number in the vertical direction of Fig. 5.13a. The *solid line* shows the result of an evaluation by the method described in Sect. 5.3.1 and Fig. 5.9. The evaluation shows artificial oscillations that deviate from the true local amplitude of the (002) reflection, shown by the *dashed curve*. (b) Results of evaluations obtained by the following methods based on the decomposition of the image into image unit cells: computation of local Fourier coefficients (LFC, *green*) and model fitting in real space (SF, *red*). The *black crosses* show the true profile. The *orange curve* was obtained by model fitting in real space applied to the noise-reduced image shown in Fig. 5.13c. Note that the green and orange curves can hardly be seen because they are covered by the red curve [Please see Plate 10 for color reproduction of this plate]

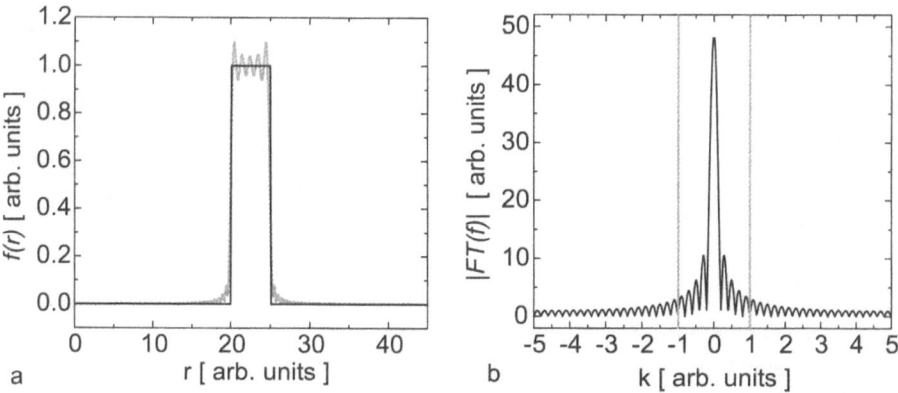

Fig. 5.15. Graphs demonstrating the effect of filtering in Fourier space. (a) The *black curve* shows the original square function $f(r)$. (b) Amplitude of the Fourier transform of $f(r)$. Only the information inside the two red lines has been used for the inverse Fourier transform. The result is the *red curve* in (a), which shows artificial oscillations due to the filtering applied in Fourier space [Please see Plate 11 for color reproduction of this plate]

$$\tilde{I}(\mathbf{r}) = \mathcal{F}\left[I_\infty(\mathbf{r})W(\mathbf{r})\right] = \tilde{I}_\infty(\mathbf{r}) \otimes \left[\mathcal{F}W(\mathbf{r})\right] . \tag{5.33}$$

The effect of the data-windowing in the diffractogram therefore corresponds to a convolution with the Fourier transform of the window function $W(\mathbf{r})$. In analogy with Fig. 5.15a, where the black curve now represents the window function, filtering in Fourier space cuts off parts of the Fourier transform of the window function $W(\mathbf{r})$ and thus yields artificial oscillations close to the image borders, as demonstrated by the red curve in Fig. 5.15a.

5.3.3.2 Decomposition of the Image into Unit Cells The following two alternative methods for the measurement of local amplitudes of reflections avoid the need for filtering in Fourier space. Both are based upon the decomposition of the image into *unit cells* (Fig. 5.13a). Since each unit cell is processed separately, it immediately becomes clear that the unit cells are not influenced by the image borders. Additionally, only those image cells that lie in the interface region between two different materials can be influenced by the transition. Thus, both origins of artificial oscillations in the measured local reflection amplitudes are avoided.

The formation of image unit cells may be performed similarly to Sect. 4.1.2, where gridding is described for ZA HRTEM images showing a dot pattern. In the present case the image pattern is different because the fringes determine only one set of grid lines. One possible method to perform the gridding here is to search for intensity maxima along equally spaced lines perpendicular to the lattice fringes (see the white lines in Fig. 5.13a). When the position of the brightness maxima have been found for each of the line scans, an image unit cell is defined by the positions of four intensity maxima at its corners, as shown in Fig. 5.13a. In this way, a grid is obtained where the shape and size of the individual unit cell depends upon the local orientation and spacing of the fringes.

A second method for gridding is the subdivision of the image into unit cells that all have the same size, orientation and shape. One set of parallel grid lines is required to be parallel to the (002) lattice fringes, and the spacing of the grid lines should be equal to the mean spacing of the (002) fringes. The required data of fringe orientation and spacing are most easily obtained by a measurement of the position of the center of the (002) reflection in the diffractogram.

5.3.3.3 Alternative Methods to Obtain Local Reflection Amplitudes The first of these alternative methods is based on a separate Fourier transform of each of the image unit cells. However, one has to take into account that the unit cells may differ from an ideal rectangular shape if the lattice fringes are bent. Therefore, each of the image unit cells is transformed into a square cell by the procedure described in Sect. 4.2.1. The square cell is subsequently Fourier-transformed and the amplitude of the appropriate Fourier coefficient is computed. A detailed description of how the Fourier coefficients are calculated in practice, as well as the resulting restrictions on the gridding used will be given in Sect. 5.3.3.5.

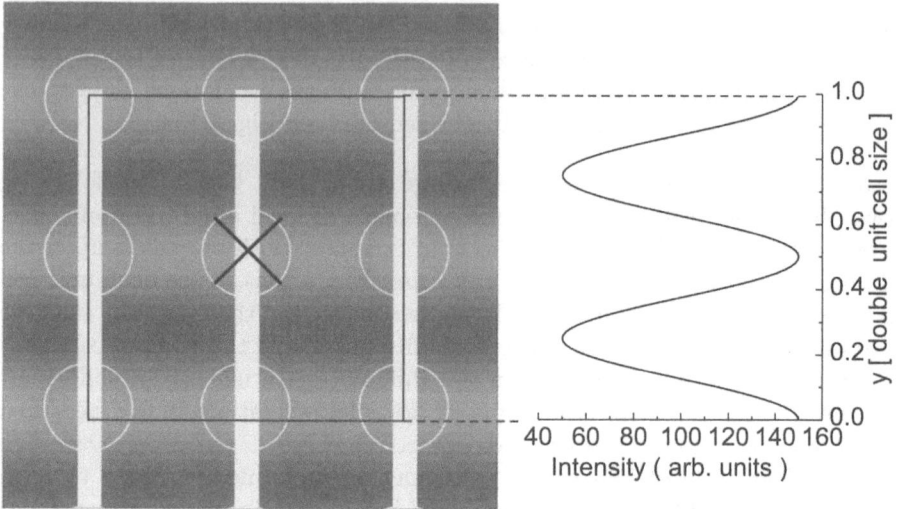

Fig. 5.16. Schematic drawing showing an area consisting of 2×2 unit cells. Averaging the intensity along the horizontal direction gives the intensity profile shown at the *right-hand side*. For the measurement of the local reflection amplitude, the intensity profile is fitted by an appropriate model and the resulting amplitude is assigned to the center position, marked by a *black cross*

The second method is based on model fitting in real space. Here we consider patches consisting of 2×2 unit cells, as shown in Fig. 5.16. Similarly to the first method, the patch is transformed into a square patch by the procedure outlined in Sect. 4.2.1. The intensity is then averaged along the direction parallel to the fringes. In this way, we obtain a profile of the image intensity like that shown at the right-hand side of Fig. 5.16. The profile is then fitted using an appropriate model. For the two-beam condition, for example, we would use a model of the form:

$$I(y) = (B + Cy) + (D + Ey) \cos \{4\pi [y + (F + Gy)]\} , \qquad (5.34)$$

where $y = 0, \ldots, 1$ is the spatial coordinate of a position within the profile. The factor $(B + Cy)$ describes the background intensity (local amplitude of the (000) reflection), $(D + Ey)$ the amplitude of the oscillation (local amplitude of the (002) reflection) and $(F + Gy)$ a phase shift. All these quantities depend linearly on y, allowing a linear variation of amplitudes and phases within the profile. The result for the amplitude of the (002) reflection, for example, given by $D + (1/2)E$, is assigned to the center position (marked by a cross in Fig. 5.16).

Both methods described above were applied to the image shown in Fig. 5.13a. The resulting profiles of the evaluated local amplitude of the (002) reflection are shown in Fig. 5.14b. The green curve was obtained by the method of calculating local Fourier coefficients (LFC), and the red curve was obtained by model fitting using (5.34). The black crosses mark the true profile. Obviously, the results of neither of the two methods show artificial oscillations, and both results resemble the true profile well.

5.3.3.4 Effect of Noise Reduction The image shown in Fig. 5.13a has been used to reveal the impact of noise reduction on the measured local reflection amplitudes. Poisson-distributed noise was added to the image as depicted in Fig. 5.13b. A noise reduction procedure using a Wiener filter was then applied as outlined in Sect. 4.1.1. The result is shown in Fig. 5.13c. An evaluation of the local amplitude of the (002) reflection from Fig. 5.13c was carried out using the model fitting method. The profile determined is shown by the orange curve in Fig. 5.14b. However, this curve is hardly visible because it is covered by the red curve, which corresponds to the evaluation of the original image. A very close inspection of Fig. 5.14b reveals small fluctuations in the GaAs region at the left-hand side of the curve. Note that the shape of the evaluated profile has not changed, which is particulary important close to the transition regions between GaAs and $In_xGa_{1-x}As$.

This consideration shows that the main effect of the noise reduction is the occurrence of (small) fluctuations. This effect can be expected because the application of the Wiener filter leads to a damping of that information whose intensity in Fourier space is comparable to the mean intensity of the noise. This effect is similar to that shown in Fig. 5.15. The oscillations of the Fourier-transformed function $f(r)$ (Fig. 5.15b) become damped if their amplitude is comparable to the local mean noise amplitude. Similarly to the red curve in Fig. 5.15a, the damping of higher spatial frequencies results in artificial oscillations and fluctuations, whereas the impact on the slope of the transition is rather small.

5.3.3.5 Homogeneous Gridding and the Discrete Fourier Transformation In the preceding description, one simple method of homogeneous gridding was mentioned that leads to image unit cells that all have the same size, shape and orientation. The size of the unit cells is given by the mean spacing of the (002) lattice fringes, and one set of parallel grid lines run parallel to the mean direction of the (002) fringes. Here we explain why this procedure leads to small errors when used in combination with the LFC method described above, if the lattice parameter of the crystal changes within the image area. This is the case for the simulated image of the $In_xGa_{1-x}As$ layer depicted in Fig. 5.13a, for example, where the (002) lattice fringes have a larger spacing in the $In_xGa_{1-x}As$ region than in the GaAs, because $In_xGa_{1-x}As$ has a larger lattice parameter than GaAs. To demonstrate this effect, we shall take a closer look at how the Fourier coefficients are computed in practice.

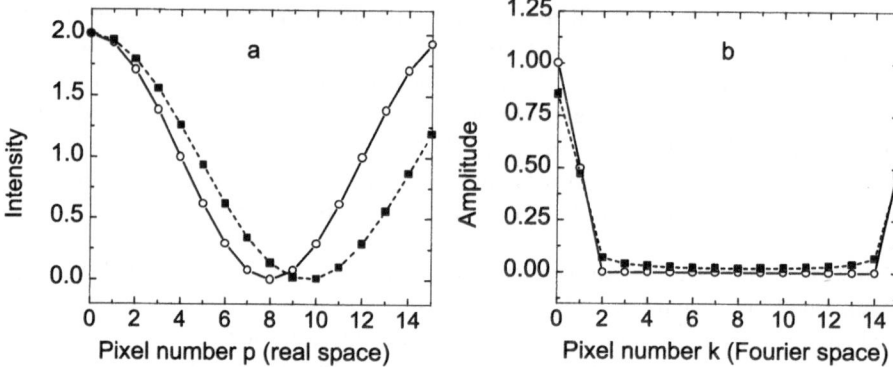

Fig. 5.17. Graphs showing the effect of a unit cell size that is different from the (002) lattice fringe spacing. (**a**) Intensity profile of a unit cell corresponding to a line scan perpendicular to the (002) lattice fringes. The line scan consists of 16 pixels. The *solid line* represents an oscillation whose period exactly fits the unit cell dimensions. The period of the *dashed oscillation* is 20% larger, but its amplitude is identical to that of the solid curve. (**b**) Amplitude of the DFTs of the pixel arrays shown in (a). The amplitude of the oscillations in (a) corresponds to (twice) the amplitude of pixel 1 of the DFT

Images evaluated with a computer are not continuous but consist of finite numbers of picture elements (pixels). As a consequence, the diffractograms of unit cells are computed by means of a *discrete Fourier transform* (DFT). The (one-dimensional) DFT of a pixel array consisting of N pixels with intensities $I(p)$ $(p = 0, \ldots, N-1)$ is given by

$$\widetilde{I}(k) = \frac{1}{N} \sum_{p=0}^{N-1} I(p) \exp \left\{ \frac{2\pi i k p}{N} \right\} , \qquad (5.35)$$

where $k = 0, \ldots, N-1$ numbers the pixels of the array $\widetilde{I}(k)$. A pixel k corresponds to the spatial frequency

$$g = \begin{cases} k/N \text{ pixel}^{-1} & \text{for } k \leq N/2 \\ -(N-k)/N \text{ pixel}^{-1} & \text{for } k > N/2 \end{cases} . \qquad (5.36)$$

The pixel $k = 1$ of the DFT corresponds to the smallest frequency $g = 1/N$ pixel^{-1} (1 oscillation period per N pixels of the real image). The largest possible frequency is found at position $k = N/2$, related to the spatial frequency $g = 1/2$ (1 oscillation period per 2 pixels). All pixels with $k > N/2$ correspond to negative frequencies. The inverse discrete Fourier transform (IDFT) is given by

$$I(p) = \sum_{k=0}^{N-1} \widetilde{I}(k) \exp \left\{ -\frac{2\pi i k p}{N} \right\} . \qquad (5.37)$$

We now consider an image unit cell chosen as shown in Fig. 5.13a. An example of an intensity profile perpendicular to the (002) lattice fringes is depicted by the solid curve in Fig. 5.17a, where a unit cell of size 16×16 pixels has been assumed. The amplitude of the DFT of the pixel array is given in Fig. 5.17b. The amplitude of the (002) reflection is found at $k = 1$, giving the result $A(g_{002}) = 0.5$.

The dashed line in Fig. 5.17a corresponds to an oscillation whose period is 20% larger than the size of the unit cell. The amplitude of its DFT is given by the dashed line in Fig. 5.17b. Using the pixel $k = 1$ for the estimation of the (002) amplitude now gives a value of $A(g_{002}) = 0.471$, which differs from the true value of 0.5 by 0.029.

This example shows that a measurement of reflection amplitudes using the DFT requires a unit cell dimension that is identical to the corresponding local spacing of the (002) lattice fringes. Choosing a homogeneous gridding leads to small errors when the unit cell dimension and the (002) lattice spacing are different.

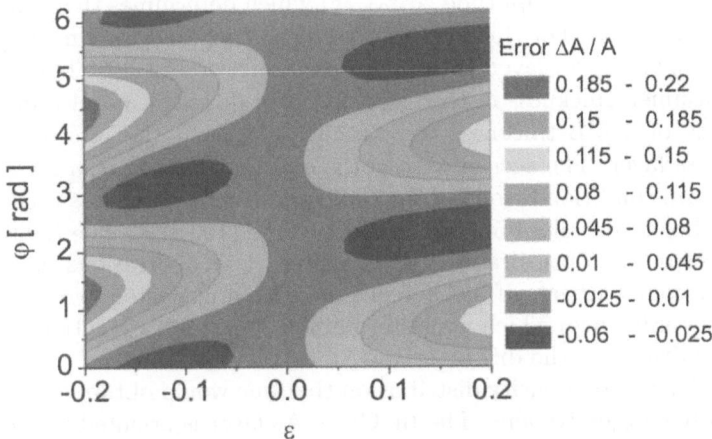

Fig. 5.18. Relative error $\Delta A(\mathbf{g})/A(\mathbf{g})$ of the measurement of the local amplitude A of a reflection $\mathbf{g}$ by the local-Fourier-coefficient (LFC) method, induced by a relative difference ϵ between the unit cell size and the period of the lattice fringe image formed by a two-beam interference of the (000) and $\mathbf{g}$ beams [Please see Plate 12 for color reproduction of this plate]

Figure 5.18 shows the error $\Delta A/A$ that arises from evaluating the amplitude of the oscillation

$$I(p) = \text{const.} + A \cos \left[2\pi p \frac{g}{N(1 + \epsilon)} + \varphi \right] \qquad (5.38)$$

by use of the DFT, where N is the number of pixels and ϵ describes the misalignment between the unit cell dimension and the spacing of the lattice fringes. The additional phase angle φ allows the possibility of a shift between the unit cell and the fringes. Note that the resulting deviation of the measured amplitude does not depend on the number of pixels N. Figure 5.18 shows that the error depends on the phase angle φ. The maximum error increases with increasing ϵ. From Fig. 5.18, we can deduce the rule of thumb that $\Delta A/A < \epsilon$. This rule applies only for a single oscillation corresponding to a two-beam condition, for example a condition involving the (000) and (002) beams. Note that the model fitting method for obtaining reflection amplitudes does not require a unit cell size that is equal to the period of the lattice fringe image. Therefore, that method can be applied with good results if a homogenous gridding is used.

The next two subsections focus on sources of error in the CELFA evaluation technique.

5.3.4 Errors Due to Objective Lens Aberrations

In this section, we consider the impact of objective lens aberrations on the evaluated profile of the local amplitude $A(\mathbf{g}_{002}, \mathbf{r})$ which determines the measured concentration profile. For this purpose, we have simulated the imaging of a 5 nm thick $In_xGa_{1-x}As$ layer with an In concentration of 30% buried in GaAs. The specimen thickness in the electron beam direction was 15 nm. Transition regions of 2 ML thickness were used for each of the interfaces of the $In_{0.3}Ga_{0.7}As$ layer. The acceleration voltage of the electron beam was 200 kV. A two-beam condition involving the (000) and (002) beams was used, and the electron beam was tilted 5° with respect to the (100) ZA. The (002) beam was parallel to the optical axis. Only coherent lens aberrations have been taken into account. Effects of incoherence have been neglected here owing to their minor influence, which consists mainly of a small reduction of the reflection amplitudes in the diffractogram.

Figure 5.19a shows the intensity distribution that one would obtain for an ideal, aberration-free objective lens. The $In_xGa_{1-x}As$ layer is oriented in the vertical direction and the (002) lattice fringes are parallel to the interfaces. For the images in Figs. 5.19b–d, a spherical-aberration constant $C_S = 1.2$ mm was used, together with defocus values $\Delta f = -200$ nm (b), -94.4 nm (c) and 0 nm (d). Figure 5.19 clearly reveals the effect of delocalization. The dark bands that are visible near the centers of the images are due to a reduced amplitude of the (000) beam in the $In_xGa_{1-x}As$ layer compared with the GaAs. In Figs. 5.19b,d, the dark band is shifted, as marked by arrows.

This effect also becomes obvious from examination of Fig. 5.20, which shows the corresponding profiles of the local amplitudes of the (000) and (002) beams. The (000) beam becomes delocalized at defocus values $\Delta f = -200$ nm and 0 nm. Note that the (000) beam is not parallel to the optical axis but is shifted in Fourier space to a position that corresponds to the spatial frequency of the (002) lattice planes, and thus is delocalized in real space in accordance with the gradient of the wave aberration, as described in Sect. 3.2.4. In Sect.

Fig. 5.19. (a) Simulated image of a 5 nm thick $In_{0.3}Ga_{0.7}As$ layer buried in GaAs. The specimen thickness in the electron beam direction is 15 nm. The transition regions from GaAs to $In_{0.3}Ga_{0.7}As$ and vice versa have a thickness of 2 ML each. The In concentration varies linearly within a transition region. The acceleration voltage is 200 kV. A two-beam condition was used in which only the (002) and (000) beams contribute to the image formation. The electron beam direction is tilted 5° with respect to the [100] ZA. The (002) beam is parallel to the optical axis. The image was calculated for an ideal objective lens without lens aberrations. The images **(b,c,d)** were computed as in (a) but taking into account spherical aberration using an aberration constant $C_S = 1.2$ mm and defocus values of **(b)** −200 nm, **(c)** −94.4 nm and **(d)** 0 nm. The *arrows* mark a dark region which corresponds to the position of the $In_{0.3}Ga_{0.7}As$ layer transferred by the (000) beam, revealing the delocalization of the information contained in the (000) and (002) beams

3.2.4 it was also shown that the delocalization of a beam vanishes if the aberration function has an extremum at the position of the beam in Fourier space. Here, this condition is fulfilled for the (000) beam at a defocus value of $\Delta f_0 = -94.4$ nm, as displayed in Fig. 3.4. Accordingly, the dark band is at its correct position in Fig. 5.19c, corresponding to the upper blue curve in Fig. 5.20. In contrast to the (000) beam, the (002) beam does not show any

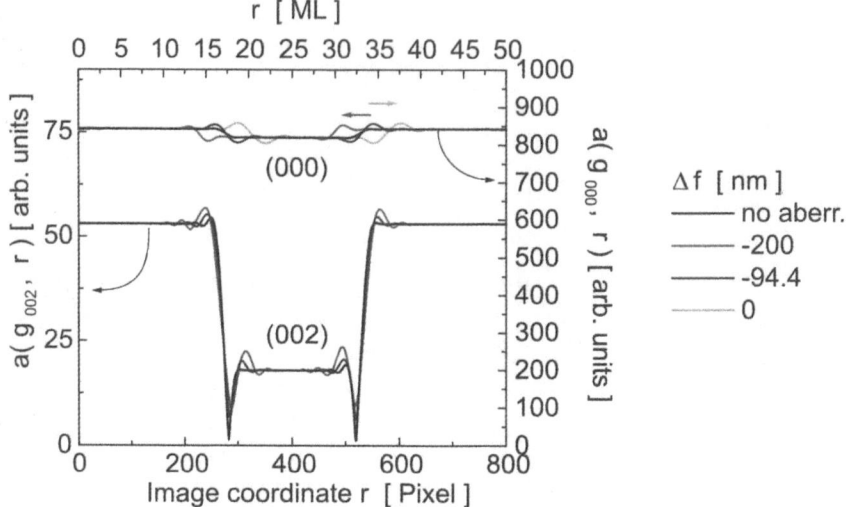

Fig. 5.20. Profiles of the local beam amplitudes $a(\mathbf{g}_{002}, \mathbf{r})$ and $a(\mathbf{g}_{000}, \mathbf{r})$ plotted versus the spatial coordinate corresponding to the horizontal direction in Fig. 5.19. The amplitude profiles are given for defocus values $\Delta f = -200$, -94.4 and 0 nm, corresponding to Figs. 5.19b, c and d, respectively. The *three colored curves* were obtained from the aberrated wave function. The *black curves* correspond to the aberration-free wave function. The *arrows* show the delocalization of the (000) beam with respect to the nondelocalized (002) beam [Please see Plate 13 for color reproduction of this plate]

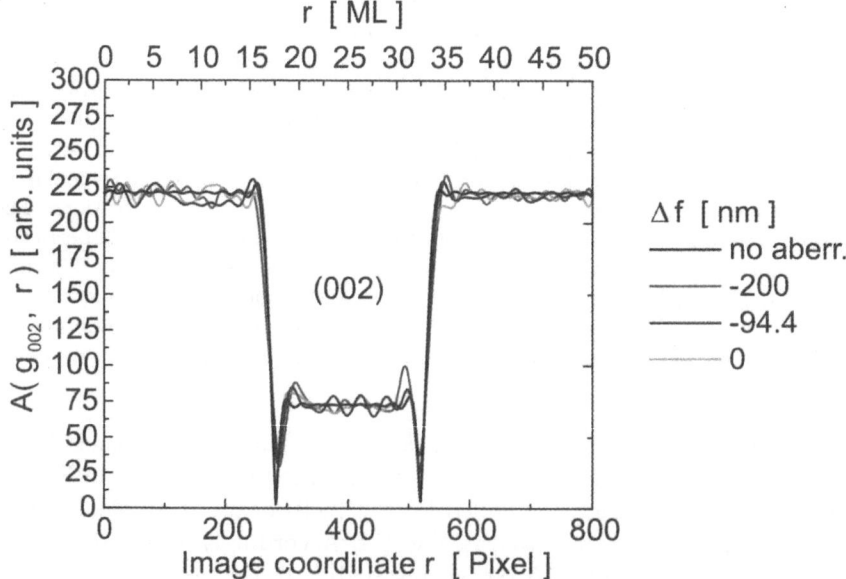

Fig. 5.21. Profiles of the local amplitude $A(\mathbf{g}_{002}, \mathbf{r})$ of the (002) reflection for various defoci, analogously to Fig. 5.20 [Please see Plate 14 for color reproduction of this plate]

effects of delocalization, owing to its position on the optical axis in Fourier space. Figure 5.20 shows that the lens aberrations cause artificial oscillations mainly in places close to the transition regions between $In_xGa_{1-x}As$ and GaAs. Figure 5.21 exhibits a similar behavior for the local amplitude of the (002) reflection, which was evaluated from the intensity distributions of the images. The lens aberrations cause small artificial oscillations for all of the defocus values shown. These considerations clearly show that the correct adjustment of the defocus by the microscope operator is not critical. However, in cases where the amplitude of the (000) reflection is more sensitive to the chemical composition, an artificial shift of the (000) beam should be avoided by choosing a delocalization-free imaging condition which is always possible for a two-beam or three-beam imaging condition in which the (002) reflection is centered on the optical axis. If, on the other hand, the chemical sensitivity of the (000) reflection is small, one should choose a defocus that minimizes the curvature of the aberration function at the optical axis.

5.3.5 Errors Due to Specimen Thickness Uncertainties

In the following, we discuss the way in which accuracy of the composition analysis depends on the error Δt in the determination of the specimen thickness in the electron beam direction. It will become obvious that the specimen thickness does not need to be known accurately.

In Sect. 5.3.1, it was pointed out that our method for composition evaluation is based on the measurement of the normalized signed modulus $A_{NS}(g_{002}, r)$, which is given by $A_{NS}(g_{002}, r) = A_S(g_{002}, r)/A_S^{Ref}(g_{002}, r)$, according to (5.24). Here $A_S(g_{002}, r)$ is the measured signed local modulus of the (002) reflection, given by (5.5), and $A_S^{Ref}(g_{002}, r)$ is the signed local reference modulus, which is either measured at a reference position $r = r_0$ in the substrate or obtained locally as described in Sect. 5.3.2. According to (5.17), $A_S(g_{002}, r)$ is proportional to the modulus $a(g_{002}, x, t)$ of the (002) beam, where we have indicated explicitly the dependence of the modulus on the composition x and the specimen thickness t. We obtain the following relation for the measured normalized signed modulus:

$$A_{NS}(g_{002}, r) \propto \frac{a(g_{002}, x, t)}{a(g_{002}, 0, t)} . \tag{5.39}$$

It is therefore instructive to take a look at the thickness dependence of $a(g_{002}, x, t)/a(g_{002}, 0, t)$, which is depicted in Fig. 5.22, plotted versus the composition x and the specimen thickness t, for $In_xGa_{1-x}As$, $Cd_xZn_{1-x}Se$ and $Al_xGa_{1-x}As$, obtained by Bloch wave calculations. From the kinematical approach to electron scattering, a linear dependence of $a(g_{002}, x, t)/a(g_{002}, 0, t)$ on the specimen thickness is expected (see e.g. Fig. 2.7). Figure 5.22 shows that the linear relationship with x is well maintained and, to a good approximation, is independent of the specimen thickness t. Only for $Al_xGa_{1-x}As$

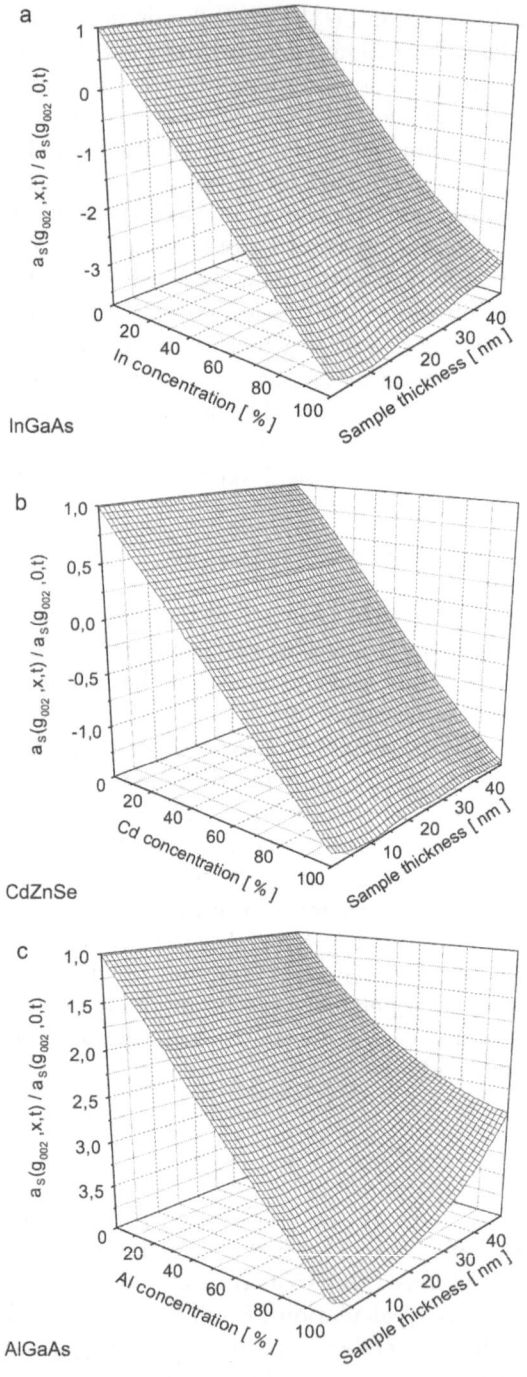

Fig. 5.22. Modulus $a(g_{002}, x, t)$ of the (002) beam normalized with respect to $a(g_{002}, 0, t)$, plotted versus the sample thickness t and the composition x for (**a**) $In_x Ga_{1-x} As$, (**b**) $Cd_x Zn_{1-x} Se$ and (**c**) $Al_x Ga_{1-x} As$

do we obtain a significant deviation, for Al concentrations larger than 50%. For that material it is thus desirable to accurately measure the local specimen thickness, which is possible by applying electron holography as shown in Chap. 9.

In the following, we show quantitative data on the impact of thickness uncertainties on composition determination in $In_xGa_{1-x}As$ and $Cd_xZn_{1-x}Se$. We have assumed an estimated thickness of 15 nm and computed the maximum deviation Δx of the composition determination for a range of the real specimen thickness t between 0.5 and 70 nm, for concentrations x between 0 and 100%. For the three-beam condition, we also take into account the fact that the phase angle φ in (5.17) may change within the image, for example because of a locally varying defocus caused by a tilt of the crystal. The error calculation was done in the following way.   Let x_{true} be the true composition at a certain position in the image under consideration and let t_{true} be the true specimen thickness. Let us assume that we have determined a local reference amplitude $A_S^{\text{Ref}}(\mathbf{g}_{002}, \mathbf{r})$ from the image. As already pointed out, the phase angle φ in (5.17) may also vary within the image. For a sufficiently slow variation, the corresponding amplitude variation is contained in the measured reference amplitude $A_S^{\text{Ref}}(\mathbf{g}_{002}, \mathbf{r})$. The measured normalized signed amplitude of the (002) reflection is thus given by

$$A_{\text{NS}}(\mathbf{g}_{002}, \mathbf{r}) = \frac{A_S(\mathbf{g}_{002}, x_{\text{true}}, t_{\text{true}}, \varphi_{\text{true}})}{A_S(\mathbf{g}_{002}, x = 0, t_{\text{true}}, \varphi_{\text{true}})} . \tag{5.40}$$

We now assume that we perform a CELFA evaluation with a wrong set of data concerning the specimen thickness and the phase angle φ. Since only an image with clearly visible (002) lattice fringes should be evaluated, the assumption $\varphi = 0$ is meaningful because this corresponds to the maximum amplitude of the (002) reflection for a sufficiently thin specimen. We now wish to know the error Δx in the composition determination induced by the assumption of $t = 15$ nm and $\varphi = 0$ as a function of the true data $t_{\text{true}} = 0.5-70$ nm, $x_{\text{true}} = 0-100\%$ and $\varphi_{\text{true}} = 0-2\pi$. The error Δx is obtained by comparing $A_S(\mathbf{g}_{002}, x_{\text{true}}, t_{\text{true}}, \varphi_{\text{true}})/A_S(\mathbf{g}_{002}, 0, t_{\text{true}}, \varphi_{\text{true}})$ with $A_S(\mathbf{g}_{002}, x, t = 15$ nm $, \varphi = 0)/A_S(\mathbf{g}_{002}, 0, t = 15$ nm $, \varphi = 0)$ and finding the concentration x with the best match, which call x_{best}. The error is then given by $\Delta x = x_{\text{best}} - x_{\text{true}}$. To be able to print the result as a graph, φ_{true} was varied from 0 to 2π for each thickness and concentration, and only the maximum error Δx is given here. These calculations were also carried out for a two-beam condition, as well as for the interference of the (002) reflection with a spatially homogenous reference beam $\mathbf{g}_{000}$, which could be obtained by the application of electron holography. Of course, these two methods do not depend on the local defocus, because the dependence of (5.17) on φ vanishes if we put $a(\mathbf{g}_{004}) = 0$ in (5.17). The calculations were performed for an acceleration voltage of 200 kV and with the center of the Laue circle at $(0, 20, 2)$, corresponding to a strongly excited (004) reflection and a 5° tilt from the (100) ZA.

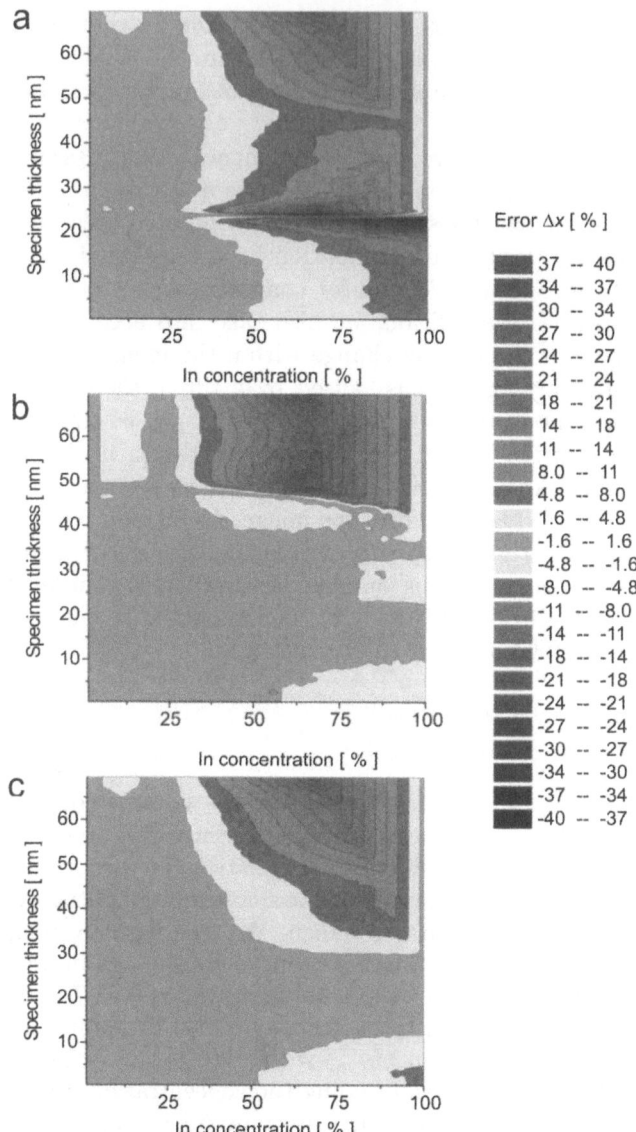

Fig. 5.23. Color-coded maps of the error Δx in a composition evaluation using CELFA for $In_xGa_{1-x}As$ induced by the assumption of a specimen thickness $t = 15$ nm and a phase angle $\varphi = 0$ (see (5.17)), plotted versus the true concentration x_{true} and the true specimen thickness t_{true}, for (**a**) three-beam condition, (**b**) a two-beam condition and (**c**) a two-beam interference of the (002) beam with a spatially homogeneous (000) reference wave, as obtained for the (002) reflection using electron holography. The legend gives the error Δx in units of percent. A detailed description of the imaging conditions is given in the text [Please see Plate 15 for color reproduction of this plate]

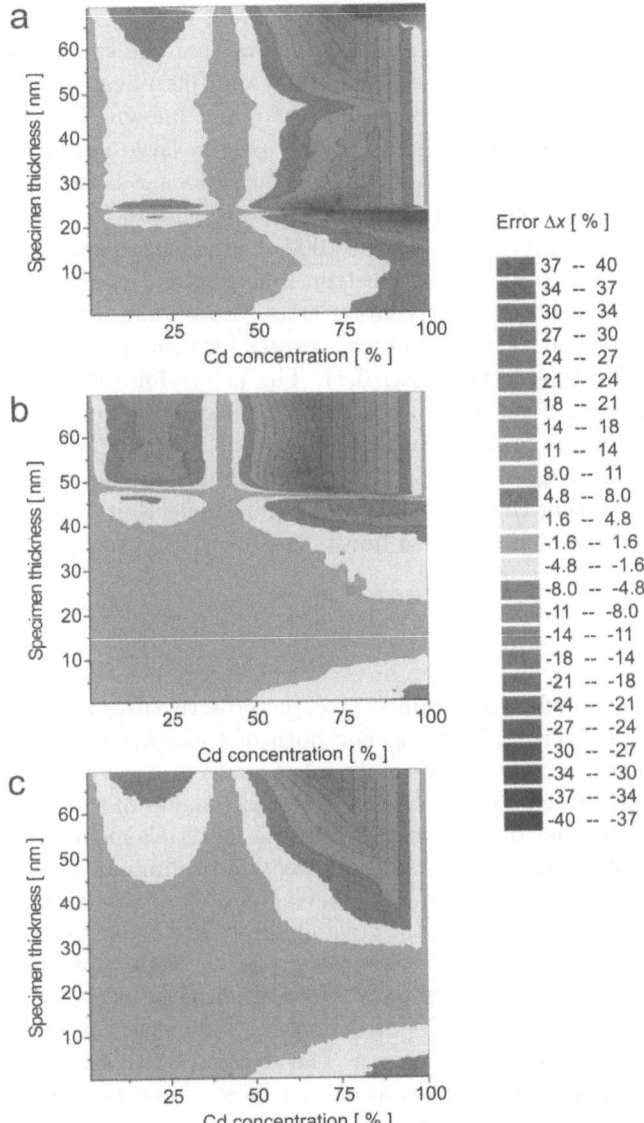

Fig. 5.24. Color-coded maps of the error Δx in a composition evaluation using CELFA for $Cd_xZn_{1-x}Se$ induced by the assumption of a specimen thickness $t = 15$ nm and a phase angle $\varphi = 0$ (see (5.17)), plotted versus the true concentration x_{true} and the true specimen thickness t_{true}, for (**a**) three-beam condition, (**b**) a two-beam condition and (**c**) a two-beam interference of the (002) beam with a spatially homogeneous (000) reference wave, as obtained for the (002) reflection using electron holography. The legend gives the error Δx in units of percent. A detailed description of the imaging conditions is given in the text [Please see Plate 16 for color reproduction of this plate]

The results are given as color-coded maps in Fig. 5.23 for $In_xGa_{1-x}As$ and Fig. 5.24 for $Cd_xZn_{1-x}Se$. Figures 5.23b and 5.24b show that the two-beam condition exploiting the interference between the (002) and (000) beams gives the most accurate results if the specimen thickness t is not known. The assumption of $t = 15$ nm causes an error $\Delta x < 5\%$ over a large specimen thickness range, of nearly 0 to 40 nm, and for concentrations x between 0 and 100%. The error is particulary small for concentrations below 50%, where $\Delta x < 2\%$. Surprisingly, the interference of the (002) beam with a spatially homogenous reference beam (obtained by electron holography), as depicted in Figs. 5.23c and 5.24c, leads to a slightly larger error for an unknown specimen thickness. Obviously, the (000) beam compensates for some part of the thickness dependence of $a(\mathbf{g}_{002}, x, t)/a(\mathbf{g}_{002}, 0, t)$. The errors for the three-beam condition are given in Figs. 5.23a and 5.24a. This imaging condition leads to increased errors close to a specimen thickness of 25 nm. It can be deduced from Fig. 5.6 that $a(\mathbf{g}_{000})$ and $a(\mathbf{g}_{004})$ become similar at that thickness, with the consequence that the amplitude of the (002) reflection vanishes for a phase angle of $\varphi = 0$, which of course boosts the error.

5.4 Strain Effects

This book is mainly concerned with strained-layer heterostructures such as $In_xGa_{1-x}As$/GaAs and $Cd_xZn_{1-x}Se$/ZnSe. For both of those systems, the lattice parameter mismatch of the binary materials ($x = 0$ or 1) is close to 7%. In pseudomorphically grown structures, the lattice parameter of the layer parallel to the interface fits that of the substrate. In the materials mentioned above, the compression of the layer in the lateral direction results in a dilatation perpendicular to the interface plane. This tetragonal distortion results in a lattice parameter difference between the strained layer and the substrate that is by approximately a factor of two with respect to the bulk materials increased, as explained in Sect. 4.3.1 (see (4.26)). Figure 5.25 visualizes this situation. The specimen is thin in the [100] direction and "infinitely" large to the right and left, but only a thin slice is shown. The electron beam direction is close to [100]. The strained layer bulges at the surfaces of the sample so that the layer expands along the [100] direction. This surface effect thus leads to a reduction of the tetragonal distortion (see (4.26)) that takes place in the vertical direction in Fig. 5.25. The figure shows that the lattice parameter of the (002) planes is, consequently, different inside the strained layer and in the matrix. In addition, the (002) lattice planes are bent close to the strained-layer interfaces owing to the surface effects. This consideration clearly shows that the imaging of the (002) lattice planes is affected by strain. On the other hand, the (020) lattice planes perpendicular to the interfaces are not affected.

Figure 5.26 provides evidence that strain effects indeed influence the amplitudes of beams. This figure shows an $In_xGa_{1-x}As$ SK layer buried in GaAs. Figure 5.26a was taken by tilting the specimen by about 4° from the [100] ZA

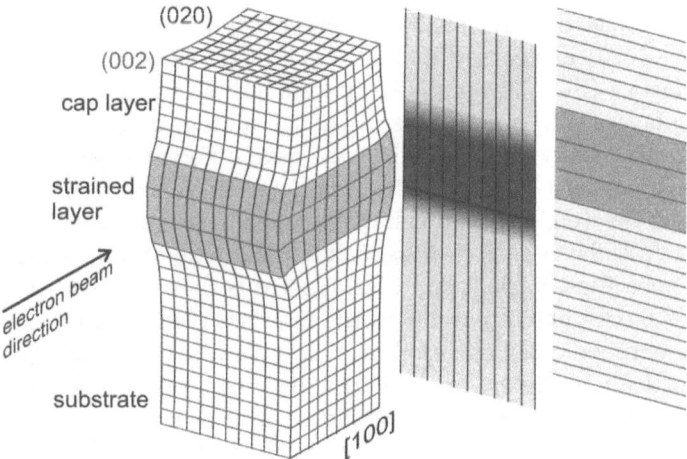

Fig. 5.25. Schematic drawing explaining the effect of strain on the imaging of the (002) and (020) lattice planes for a pseudomorphically grown strained-layer heterostructure. The tetragonal distortion and its partial relaxation due to the TEM specimen being thin along the [100] direction leads to (002) lattice planes that have different lattice parameters in the strained layer and the substrate and show lattice plane bending close to the interfaces between the strained layer and the matrix. The (020) lattice planes are not distorted. Imaging using off-ZA imaging conditions leads to a broadening of the interfaces for the (020) lattice planes and the occurrence of strain effects for the (002) planes [Please see Plate 17 for color reproduction of this plate]

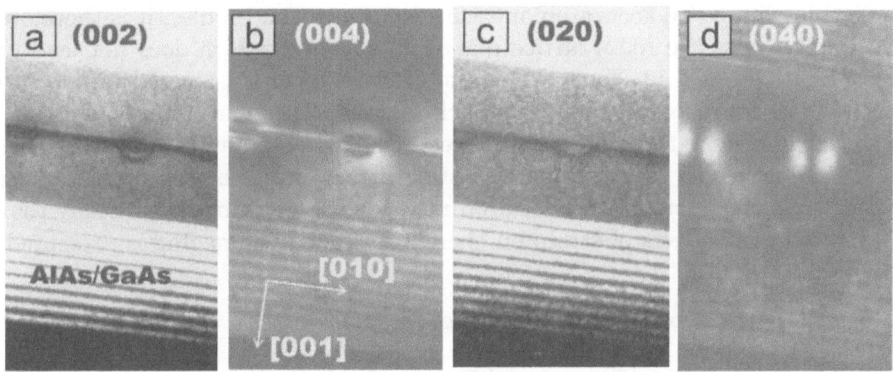

Fig. 5.26. Cross-sectional TEM DF images of an $In_xGa_{1-x}As$ SK layer buried in GaAs, revealing the effect of strain. The growth direction is [001]. The images were taken with (**a**) the (002), (**b**) the (004), (**c**) the (020) and (**d**) the (040) beam

by rotation around an axis parallel to the [001] direction. The objective aperture was centered around the (002) beam (with zero excitation error), which was aligned parallel to the optical axis. Figure 5.26b was taken by shifting the aperture to the (004) reflection. An analogous procedure was used to obtain Figs. 5.26c,d, where (020) is strongly excited. Each of the images shows an AlAs/GaAs superlattice (SL), 30 nm GaAs spacers and an InAs layer consisting of a wetting layer (WL) and islands. Although all four images display the same specimen area, they look different and have different information contents. The image contrast in Figs. 5.26a,c is determined mainly by the composition because the {002} reflections are chemically sensitive, whereas the contrast in the InAs layer seen in Figs. 5.26b,d is mainly due to strain. The difference between Figs. 5.26a,b and Figs. 5.26c,d is that the first set of micrographs images lattice planes parallel to the interface plane and the second set images the perpendicular planes. The planes parallel to the interface possess different lattice parameters in the GaAs matrix and the InAs layer. Moreover, they are bent in the vicinity of the InAs layer owing to the small thickness of the TEM specimen. Therefore, Fig. 5.26b shows a complex contrast pattern in the InAs region. The WL is not visible in Fig. 5.26d, because the lattice planes perpendicular to the interface plane are not influenced by the strain field of a homogeneous, pseudomorphically grown 2D layer. Image contrast due to lattice bending appears only at the borders of the islands.

To avoid effects of strain in composition evaluation, one should therefore choose the (020) reflection for chemically sensitive imaging. This approach, unfortunately, leads to a broadening of the interfaces in the image. If one is aiming at off-ZA imaging conditions for a CELFA analysis, the electron beam has to be tilted a few degrees around the [100] ZA. Imaging with (020) lattice planes requires the direction of the electron beam to be inclined relative to the interface plane, which leads to a blurring of the interfaces, as shown in Fig. 5.25 (blue lines). To accurately measure concentration profiles, it is therefore desirable to use the (002) lattice planes for imaging, which does not lead to broadening of the interfaces when the electron beam is aligned parallel to the (001) crystal plane. This approach, on the other hand, needs a precise knowledge of how the difference between the lattice parameters of the layer and substrate and the bending of lattice planes influence the excitation of the chemically sensitive (002) beam. In the following, a simulation study that illuminates the effects of strain on chemically sensitive imaging is presented.

5.4.1 Variation of Lattice Parameter

We first discuss the effect of a difference between the lattice parameters in the strained layer and the substrate without taking into account lattice plane bending. As described earlier, composition evaluation by CELFA is based upon a comparison of the measured local amplitudes of the (002) reflection with those obtained by Bloch wave calculations. In the case where the local lattice parameter of the structure is basically unknown, an initial CELFA analysis could rely on theoretical values of the local amplitude of the (002) reflection $A_{\mathrm{NS}}^{\mathrm{Bloch}}(\mathbf{g}_{002})$ provided by Bloch wave calculations performed

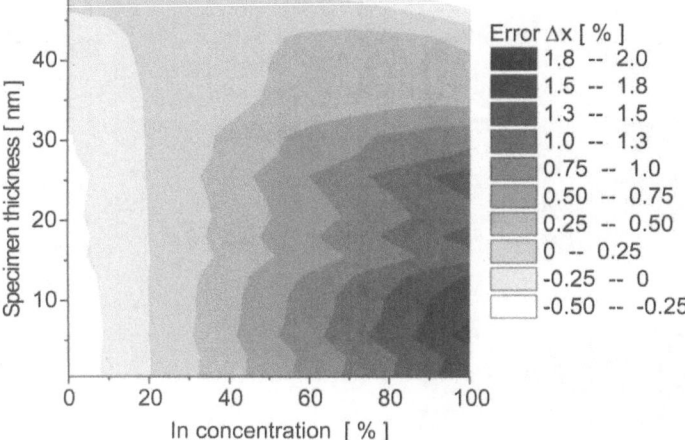

Fig. 5.27. Grayscale-coded map showing the effect of assuming an incorrect lattice parameter on the In concentration evaluated using the CELFA method, calculated for In$_x$Ga$_{1-x}$As, standard imaging conditions of the CELFA method with a strongly excited (004) reflection, and an acceleration voltage of 200 kV. The error for a 1% lattice parameter deviation is plotted versus the In concentration x and the specimen thickness. This result suggests that one should use the correct lattice parameter corresponding to the composition x when calculating the Bloch wave data that is used in the CELFA method

with the lattice parameter of the substrate. Figure 5.27 depicts a grayscale-coded graph of the associated error Δx in the composition determination for In$_x$Ga$_{1-x}$As. This plot gives the error Δx in units of percent, calculated for a deviation of 1% between the true lattice parameter and that assumed in the evaluation. Note that Fig. 5.27 does not take into account any correlation between the In concentration and the bulk material lattice parameter. Figure 5.27 indicates that the maximum error Δx due to a lattice parameter fluctuation of 1% is 2%. Assuming, for example, a pseudomorphically strained In$_x$Ga$_{1-x}$As layer with $x = 60\%$ buried in GaAs and a specimen thickness of 15 nm, and assuming a tetragonal distortion corresponding to $\alpha_R = 1.5$ (see (4.26)), one obtains a deviation of the measured In concentration of $\Delta x = 1(\%$ error per% strain$) \times 7(\%$ strain of binary material$) \times 0.6 \times 1.5 = 6.3\%$, corresponding to $\Delta x/x = 10\%$. This error can be significantly decreased if the correct lattice parameter (corresponding to $\alpha_R = 1.5$, for example) is used in the Bloch wave calculation used to obtain $A_{NS}^{Bloch}(\mathbf{g}_{002})$. Note that the correlation between the lattice parameter and the composition has been taken into account in all the evaluations of experimental data presented in Part III where the (002) reflection is exploited.

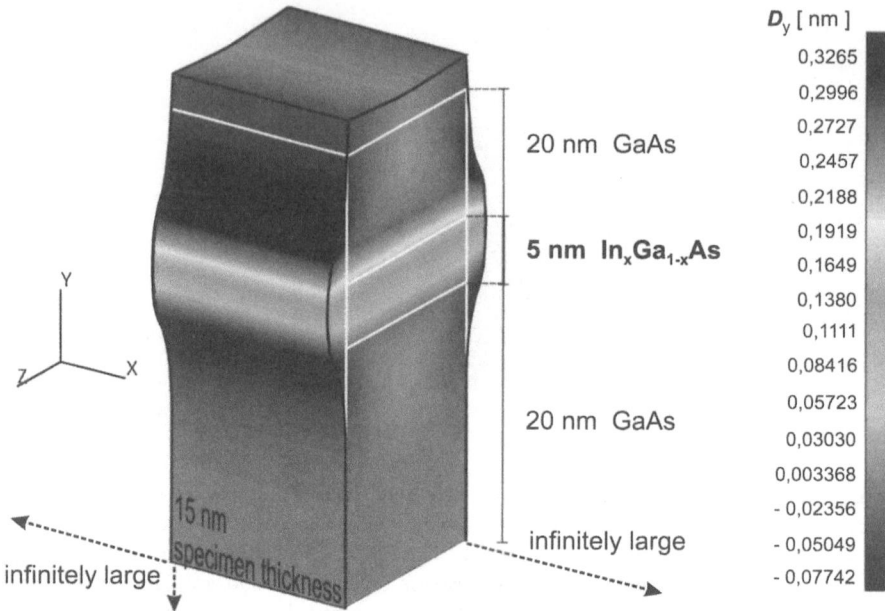

Fig. 5.28. The *white lines* show the specimen model used for the FE calculation of the strain field of a 5 nm thick $In_xGa_{1-x}As$ layer grown on a GaAs substrate and covered with 10 nm GaAs. The specimen thickness is 15 nm. The *dashed arrows* indicate directions in which the model is assumed infinitely large; this was simulated by using appropriate boundary conditions. The distorted model shows the result of an FE calculation carried out for an In concentration of $x = 50\%$. For better visibility of the deformation field, the deformation is multiplied by factor of 10. The color code corresponds to the component of the displacement vector in the growth direction [Please see Plate 18 for color reproduction of this plate]

5.4.2 Method for Simulation of Strain Effects

A study of the effects of both lattice parameter fluctuations and lattice plane bending was performed using FE simulations, as explained in Sect. 4.3.2. The FE model is indicated by the white lines in Fig. 5.28. The model consists of a specimen with a thickness of 15 nm in the electron beam direction, containing a 5 nm thick $In_xGa_{1-x}As$ layer on top of a GaAs substrate and covered with a 10 nm thick GaAs cap layer. The model is assumed infinitely large along the directions indicated by dashed arrows. The result of the FE calculation is also given in Fig. 5.28, where the deformation of the model (magnified by a factor of 10) and the color-coded component of the displacement in the [001] direction are shown.

For the simulation of the electron wave function passing through the deformed model, we used the multi slice (MS) method provided by the EMS

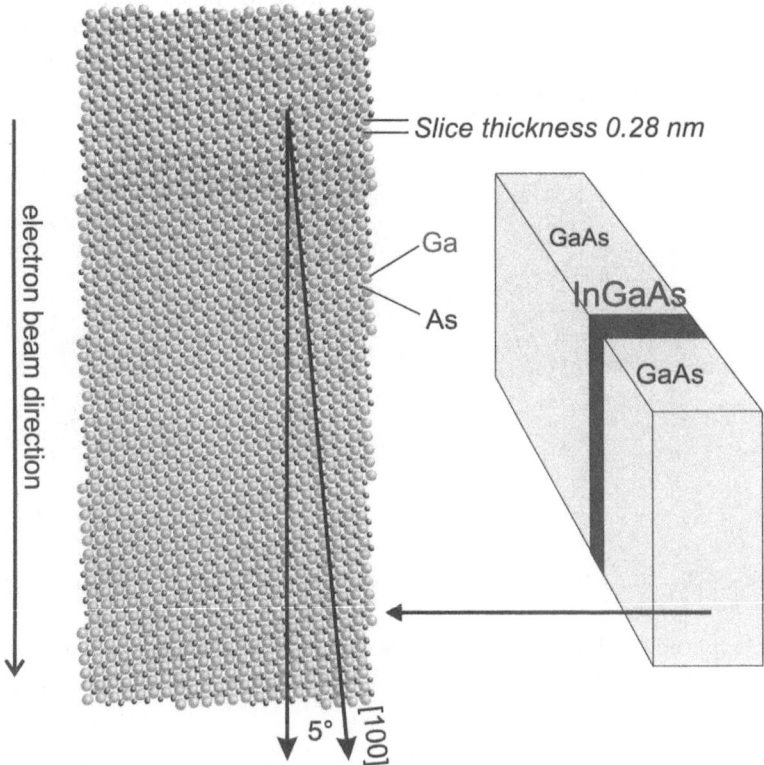

Fig. 5.29. The image on the *left* shows a projection of the 3D atomic model used for the MS simulation, corresponding to the body shown on the *right* viewed looking towards the front face. This image depicts the geometry of the FE model, which is oriented in such a way that the electron beam direction is vertical. The atomic model clearly shows the [100] crystal ZA, which is inclined at 5° from the electron beam direction

program package [2], because the Bloch wave method can be applied only in the case of a perfectly homogeneous crystal specimen. The MS method can be used to simulate electron diffraction in crystals containing different materials, strain, interfaces, or crystal defects such as dislocations and stacking faults. The MS approach is based upon dividing an atomic model of the real specimen into slices a few tenths of a nanometer thick in the electron beam direction. The interaction of the electrons with matter is simulated by a stepwise propagation of the electron wave from slice to slice. First, the wave function of the electron beam at the entrance plane of a slice is multiplied by the object transfer function computed from the projected potential of the slice. Then, the electron wave propagates freely to the entrance plane of the next slice. This procedure is continued to the exit plane of the last slice, where the wave function at the object exit surface is obtained. Figure 5.29

shows the orientation of the FE model with respect to the electron beam direction and the projection of the 3D atomic model of the deformed specimen along a direction perpendicular to the electron beam direction. The model is subdivided into slices of thickness 0.28 nm.

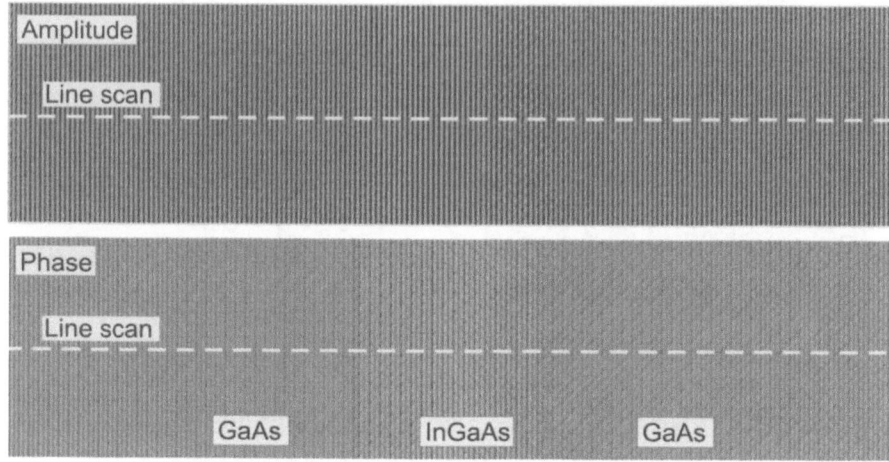

Fig. 5.30. Amplitude and phase images of the electron exit wave function obtained by the MS method, applied to the atomic model depicted in Fig. 5.29

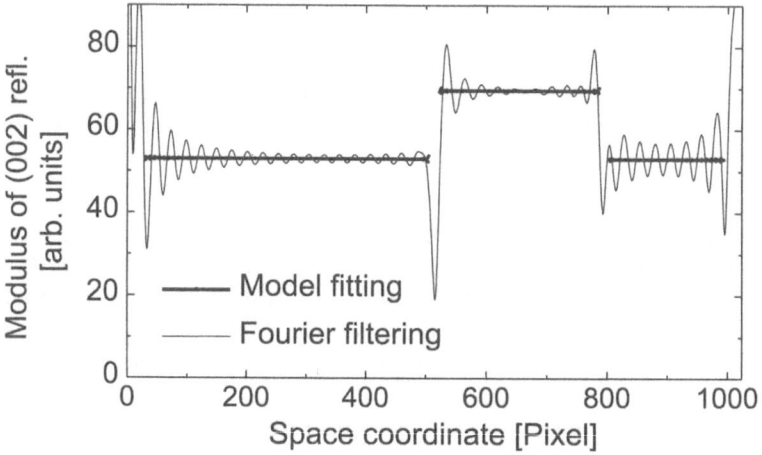

Fig. 5.31. Profile of the local modulus of the (002) reflection evaluated from the simulated wave function shown in Fig. 5.30. The *thin curve* gives the result for the Fourier filtering technique, containing strong artificial oscillations. The *thick curve* was obtained by model fitting, as explained in the text

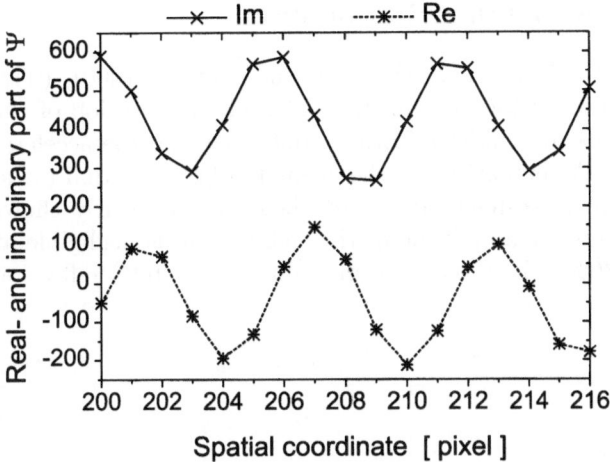

Fig. 5.32. Example of model fitting applied to a small part of the line scan marked by dashed lines in Fig. 5.30. The local fit displayed here was used to obtain the beam amplitudes at the position $n = 208$. The width of the fitted region is $\Delta n = 16$ pixels. The real and imaginary parts of the wave function were fitted simultaneously

Figure 5.30 depicts the amplitude and phase images of the resulting electron exit wave function. The next task is the evaluation of the local amplitude of the (002) reflection. As shown in Fig. 5.31, the Fourier filtering method described in Sect. 5.3.1 leads to artificial oscillations that are too strong for a quantitative analysis of strain effects. To avoid these oscillations, a model fitting approach analogous to that described in Sect. 5.3.3 was used. A line scan of the wave function Ψ_i, where $i = 1, \ldots, 1024$ was extracted, as marked by dashed white lines in Fig. 5.30. For each pixel n, a region $i \in [n - \Delta n/2, n + \Delta n/2]$, with Δn corresponding to approximately 0.3 nm, was fitted by

$$
\begin{aligned}
\Re\left[\Psi(i)\right] =\ & a(g_{000}, n)\cos\{p(g_{000}, n)\} \\
& + a(g_{002}, n)\cos\{p(g_{002}, n) + g_{002}\left[i + (n - i)\epsilon(n)\right]\} \\
& + a(g_{004}, n)\cos\{p(g_{004}, n) + 2g_{002}\left[i + (n - i)\epsilon(n)\right]\} \ ,
\end{aligned}
$$

$$
\begin{aligned}
\Im\left[\Psi(i)\right] =\ & a(g_{000}, n)\sin\{p(g_{000}, n)\} \\
& + a(g_{002}, n)\sin\{p(g_{002}, n) + g_{002}\left[i + (n - i)\epsilon(n)\right]\} \\
& + a(g_{004}, n)\sin\{p(g_{004}, n) + 2g_{002}\left[i + (n - i)\epsilon(n)\right]\} \ , \quad (5.41)
\end{aligned}
$$

where $a(g_{00j}, n)$ and $p(g_{00j}, n)$ are the modulus and phase of the (00j) beam corresponding to the pixel n, $\epsilon(n)$ is the strain, and g_{002} is the spatial frequency corresponding to the (002) planes in the unstrained substrate. The parameters $a(g_{00j}, n)$, $p(g_{00j}, n)$ and $\epsilon(n)$ were used as fitting parameters. Figure 5.32 gives an example that shows the model-fitting result around the position $n = 208$ obtained using $\Delta n = 16$.

5.4.3 Beam Amplitudes and the Effect of Strain

The results for the effect of strain on the amplitudes of the (000), (002) and (004) beams are collected together in Fig. 5.33. The orientation of the electron beam corresponds to standard CELFA conditions and the acceleration voltage was 300 kV. A comparison of the evaluated beam moduli (solid curves) with those of an unstrained sample (dashed curves) clearly shows that the effect of strain on the amplitude of the (002) beam is negligible at an In concentration of 30%, where only the modulus of the (004) reflection

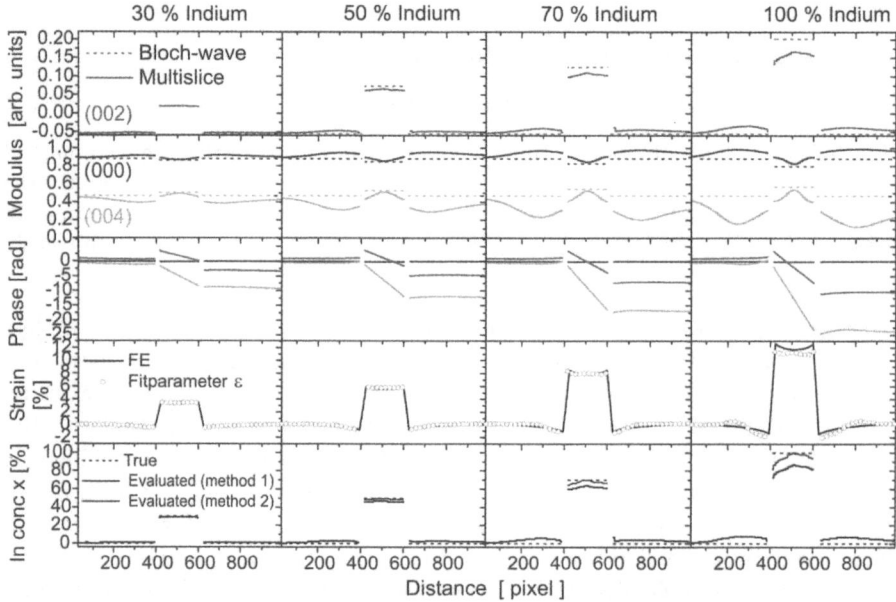

Fig. 5.33. Graph showing the effect of strain on beam amplitudes and on the evaluated In concentration x for a 5 nm thick $In_xGa_{1-x}As$ quantum well buried in GaAs. Each column of graphs corresponds to one simulation, carried out for $x = 30, 50, 70$ and 100% indium. The *upper three rows* contain the moduli and phases (*solid lines*) of the (002), (000) and (004) beams, as evaluated from the simulated wave functions. The *dashed lines* show the beam moduli obtained from Bloch wave calculations, i.e. corresponding to an unstrained specimen. The *fourth row* gives the results obtained for the strain parameter ϵ of (5.41). The *solid curves* correspond to the strain obtained directly from the FE model by averaging along the electron beam direction. The *bottom row* shows the true concentration profile (*dashed curves*) and the concentrations evaluated from the amplitude of the (002) beam with (method 1) a set of Bloch wave data calculated using the lattice parameter of GaAs and (method 2) a set of data calculated using a lattice parameter corresponding to a pseudomorphically strained layer, where a full relaxation of the tetragonal distortion was assumed (thin-sample limit) [Please see Plate 19 for color reproduction of this plate]

exhibits a considerable deviation. The deviation of the (002) modulus is still small at an In concentration of 50%, but becomes significant at $x = 70$ and 100%. The largest impact is seen in the case of the modulus of the (004) reflection, which is consistent with the strong contrast in the experimental image shown in Figs. 5.26b,d. The lower row of graphs in Fig. 5.33 shows the error in the In concentration induced by strain. These graphs were obtained by comparing the modulus of the (002) beam evaluated from the MS wave function with a set of Bloch wave data $a_S^{Bloch}(\mathbf{g}_{002}, x, t = 15 \text{ nm})$, calculated (method 1) with the lattice parameter of the substrate and (method 2) with a lattice parameter that depends on the In concentration x in a way that corresponds to a pseudomorphically grown sample assuming full relaxation of the tetragonal distortion (thin-sample limit). Figure 5.33 clearly shows that method 2 decreases the evaluation error considerably, as was already expected in Sect. 5.4.1. This figure shows that the maximum error Δx of the CELFA analysis at an In concentration of $x = 50\%$ is 4% for method 1 and 2% for method 2.

In addition to the main result of the impact of strain on the composition evaluation, Fig. 5.33 contains the phases of the three beams involved. One can clearly see the phase shift of π that occurs for the (002) reflection (red curve) at the interfaces, and a slope of the phase connected with the lattice parameter being different inside the strained layer. The phase difference that is related to the lattice parameter mismatch between the layer and substrate is called the *geometric phase*. It also could be used for the compositional analysis. The blue line corresponding to the phase of the (000) beam does not show geometric phase components, which makes it useful for the determination of the specimen thickness by electron holography. Finally, Fig. 5.33 reveals the fit-parameter ϵ which reasonably reproduces the actual strain deduced from the FE model by averaging along the electron beam direction.

5.5 Nonrandom Distribution of Elements

In the preceding section it was assumed that the electron beam, while traveling through a specimen consisting of a material $A_x B_{1-x} C$, encounters elements A and B, that occupy the same sublattice with a distribution that is random along the propagation direction. Figure 5.34 clearly reveals that the situation is different if the interface plane is inclined relative to the electron beam direction. An imaging condition like this might be chosen if strain effects have to be avoided, so that the (020) lattice planes perpendicular to the interface planes are used for the chemically sensitive imaging. Figure 5.34 also shows the In concentration profile obtained by averaging along the electron beam direction. Owing to the crystal tilt, the profile shows a slope at the interfaces between the $In_x Ga_{1-x} As$ layer and the GaAs.

In the following, the results of an MS simulation based upon the model depicted in Fig. 5.34 are described. The electron exit wave functions obtained

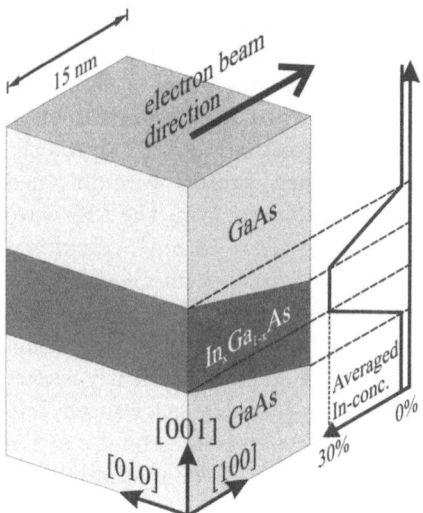

Fig. 5.34. Schematic drawing showing a nonrandom distribution of atoms along the electron beam direction due to the interfaces between the $In_xGa_{1-x}As$ layer and the GaAs being inclined with respect to the electron beam direction. The drawing also shows a profile of the In concentration averaged along the electron beam direction. The model depicted here was used to simulate the effect of a nonrandom distribution of elements on the CELFA evaluation

were evaluated as described in the previous section. Figure 5.35a depicts the concentration profiles obtained for $In_xGa_{1-x}As$, where four layers with In concentrations of 20, 30, 50 and 100% were evaluated. The true concentration profiles are indicated by solid lines. Significant deviations between the true and the evaluated profiles develop only at In concentrations close to $x_0 = 22\%$, which is marked by a horizontal dashed line. Note that x_0 corresponds to the In concentration at which the amplitude of the (002) beam vanishes. Around $x_0 = 22\%$, a "gap" is visible in the evaluated concentration. Figure 5.35b exhibits a similar behavior for $Cd_xZn_{1-x}Se$. A "gap" in the evaluated Cd concentration is now seen at $x_0 = 42\%$, corresponding to the Cd concentration at which the (002) beam vanishes.

Figure 5.36 allows a closer examination of this effect. This figure shows the modulus and phase of the (002) beam close to the interface that is inclined with respect to the electron beam direction. The dashed curves show the result of Bloch wave calculations based upon the mean concentration averaged along the beam, thus corresponding to the case of a random distribution of atoms. The most conspicuous feature is the fact that the modulus $a(\mathbf{g}_{002}) > 0$ at the position $x_0 = 22\%$, where the dashed curve is zero, which obviously leads to the "gap" in the evaluated In concentration seen in Fig. 5.35a close to $x_0 = 22\%$. Additionally, the phase of the (002) beam does not show a sharp

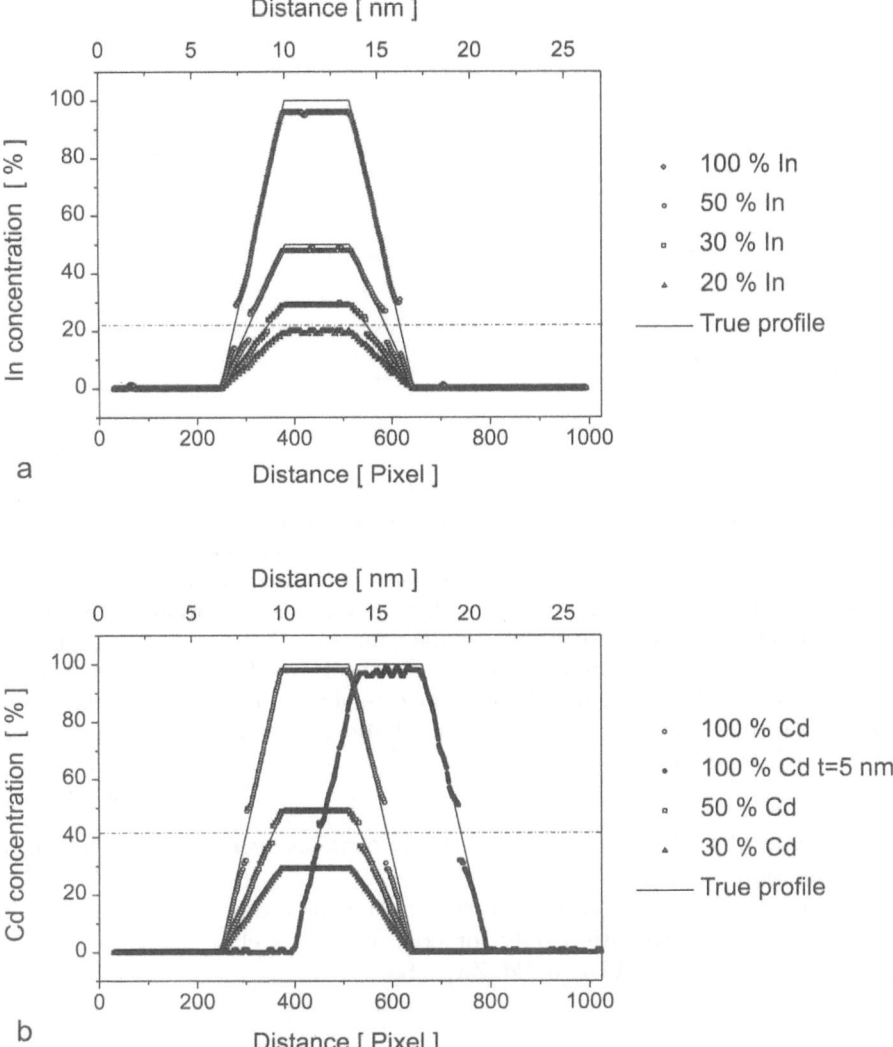

Fig. 5.35. Concentration profiles of specimens where the interfaces are inclined with respect to the electron beam direction as shown in Fig. 5.34. The profiles were obtained by evaluating electron exit wave functions that were simulated with the MS method. (**a**) Results for $In_xGa_{1-x}As$ and (**b**) results for $Cd_xZn_{1-x}Se$. The specimen thickness was 15 nm if not stated otherwise. For both material systems, "gaps" occur at the concentrations x_0 that correspond to a vanishing modulus of the (002) beam

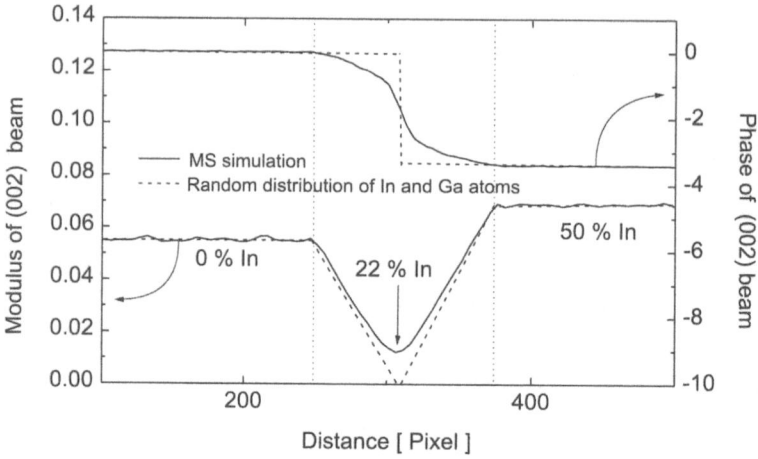

Fig. 5.36. Modulus and phase of the (002) beam close to an interface that is inclined with respect to the electron beam direction, plotted versus the distance in the [001] direction. The *dashed lines* correspond to the concentration profile shown in Fig. 5.34, but with the In and Ga atoms randomly distributed. The *solid curves* show the results for the nonrandom distribution of In and Ga atoms shown in Fig. 5.34, obtained from the evaluation of a wave function that was generated by the MS method. The nonrandom atom distribution leads to a broadening of the phase shift of π at $x_0 = 22\%$, which is sharp in the case of the random atom distribution. In contrast to the case for the random distribution, we observe that the modulus of the (002) beam does not vanish at x_0

phase jump of π at x_0, but changes slowly changes over the whole width of the transition region.

In conclusion, the nonrandom distribution of atoms leads to errors in the evaluated concentration only for concentrations close to $x_0 = 22\%$ in $In_x Ga_{1-x} As$ and $x_0 = 42\%$ in $Cd_x Zn_{1-x} Se$.

References

1. D.B. Williams, C.B. Carter: *Transmission Electron Microscopy*, 1st edn. (Plenum, New York, 1996)
2. P.A. Stadelmann: Ultramicroscopy **51**, 131 (1987)

Part III

Applications

6 Introduction

Part III presents some examples of the application of the evaluation techniques described in the previous chapters. These examples are concerned with the growth of III–V semiconductor heterostructures such as $In_xGa_{1-x}As/$ GaAs(001) and $Al_xGa_{1-x}As/GaAs$(001). The investigations demonstrate that segregation is one of the main effects influencing the growth of the III–V (and also II–VI) semiconductors. After a brief introduction to the growth of semiconductor quantum dot (QD) islands in the SK growth mode, a chronologically arranged survey of the research work that has led to the understanding of segregation effects that we presently have is given in the present chapter. We then turn to the investigation of $In_{0.6}Ga_{0.4}As$ layers buried in GaAs in Chap. 7, where we show that, besides segregation, kinetic effects also play a significant role in the growth of capped SK layers. The subsequent subject is the growth of nominally binary InAs SK islands, also buried in GaAs, in Chap. 8. We shall see that the islands contain a large amount of Ga atoms that originate from the GaAs substrate. The effect of mass transport from the GaAs substrate into the InAs islands will be discussed with respect to segregation. The last example given in Chap. 9 concerns the $Al_xGa_{1-x}As$ material system. From the point of view of compositional analysis, $Al_xGa_{1-x}As$ is a difficult system because it requires a quite accurate knowledge of the specimen thickness, as pointed out in Sect. 5.3.5 (see Fig. 5.22). It will be shown that electron holography is a valuable tool for simultaneously obtaining the local specimen thickness and the modulus of the chemically sensitive (002) reflection.

6.1 Stranski–Krastanov Growth Mode

Low-dimensional semiconductor heterostructures are one of the main research topics in solid-state physics at present. Most applications of semiconductor nanostructures are found in the field of optoelectronic devices such as light-emitting diodes and lasers. The development of QD lasers is expected to lead to an increased quantum efficiency and lower threshold current densities [1]. In some high-lattice-mismatch heterostructures such as InGaAs/GaAs, the SK growth mode is observed (see Fig. 6.1), which leads to the self-formation

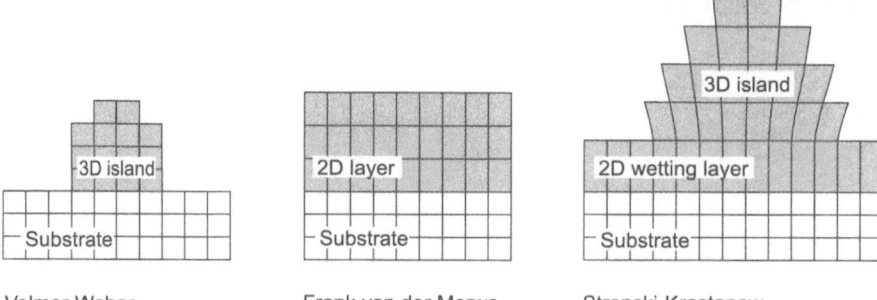

Volmer Weber   Frank van der Merwe   Stranski-Krastanow

Fig. 6.1. Schematic drawing depicting the three modes of epitaxial growth that can be distinguished. The Volmer–Weber growth mode leads to the formation of three-dimensional (3D) islands on top of the substrate. The Frank–van der Merwe growth mode results in a smooth, two-dimensional (2D) layer. The Stranski–Krastanov mode is favorable for a layer with a lattice parameter that differs considerably from that of the substrate, because the islands allow a relaxation of the strain energy. The SK growth mode leads to a 2D wetting layer and 3D islands

of QDs [2, 3]. A simplified model that explains the occurrence of island formation in the SK growth mode is based on a balance of the surface energies of the substrate and of the layer, the formation energy of the interface, the strain energy of the layer and the deformation energy of the substrate. According to this model, the SK growth mode may occur in systems where the formation of a two-dimensional layer is favorable during the deposition of the first few ML of the layer. With increasing layer thickness, the strain energy of the 2D layer increases. Above a critical thickness t_{c3D}, the onset of island formation is observed, chiefly because an island offers the possibility of relaxation of elastic strain at its free surfaces [4]. This model leads to a 2D wetting layer (WL) with 3D islands on top. The elastic relaxation of the islands is incomplete, and plastic relaxation is observed if the island size exceeds a critical value that depends on the misfit between the layer and substrate materials, the island's shape, the composition distribution inside the island, the elastic parameters, and the energy that is required for the generation of the misfit dislocations. HRTEM investigations of the strain state of free-standing $In_{0.6}Ga_{0.4}As/GaAs(001)$ and $InAs/GaAs(001)$ islands have shown [3] that the In concentration inside the islands is not homogeneous but increases from the bottom to the top of the island.

The simplified model set out above cannot be used to describe the density or size distribution of the islands. For that purpose, kinetic models of 3D island nucleation have to be applied [5]. Experimental observations carried out by Ruvimov et al. [6] with photoluminescence (PL) spectroscopy and plan-view TEM report small (12 to 14 nm) InAs islands buried in GaAs grown by MBE with an equilibrium size. These equilibrium-size islands (which are stable during growth interruptions performed prior to the cap layer growth) were observed only in a small window of the arsenic pressure of $p_0 = (2 \times 10^{-6} \pm 1 \times 10^{-6})$ Torr at a growth temperature of 480°C and a deposition of 4 ML InAs. For depositions between 2 and 3 ML InAs, equilibrium islands could be formed by use of a growth interruption with a duration between 10 s (3 ML) and 600 s (2 ML), prior to the deposition of the GaAs cap layer. At an As pressure of $p \geq 3p_0$, strain-relaxed InAs clusters appear, whereas a reduction to $p \leq 1/3p_0$ leads to the formation of macroscopic 2D islands.

The main application of InGaAs/GaAs SK structures is expected in the field of optoelectronic devices. Investigation of the local composition in buried SK layers is important from this point of view. Commonly, PL and excitation spectroscopy are applied. These methods have the disadvantage that chemical and structural effects cannot be distinguished. Structural data obtained from free-standing islands by HRTEM or atomic-force microscopy (AFM) can be used for the interpretation of the optical data only if the SK layer is not altered during the capping. However, processes such as segregation of In into the GaAs cap layer have to be expected [7].

6.2 Segregation in III–V Ternary Alloys

Segregation is a well-known effect in III–V ternary alloys such as $In_xGa_{1-x}As$ and $Al_xGa_{1-x}As$. Moison et al. [7] found a surface enrichment of one of the group III components by in-situ electron spectroscopy on structures deposited by MBE, leading to a near-binary surface. In the literature, we find two models that are used to describe the segregation in InGaAs. The first one describes the segregation in terms of an exchange reaction of atoms between only two atomic layers, the surface layer on top of the specimen and the bulk layer beneath. The exchange reaction is fast compared with a growth rate of 1 ML/s [8] and concerns one In atom in the bulk layer and one Ga atom in the surface layer that interchange their positions. Since surface and bulk sites are energetically different for both In and Ga, the chemical potentials of In and Ga are different in the bulk and surface layers. Thus, four different chemical potentials are of interest:

$$\mu_{\text{bulk(surf)}}^{\text{In(Ga)}} = \left(\frac{\partial F}{\partial N_{\text{bulk(surf)}}^{\text{In(Ga)}}} \right) , \tag{6.1}$$

where, for example, $N_{\text{surf}}^{\text{In}}$ is the number of In atoms in the surface layer and $\mu_{\text{surf}}^{\text{In}}$ is the corresponding chemical potential. In thermodynamic equilibrium, the free energy F of the system is a minimum, which leads to a balance of chemical potentials

$$\Delta F = \sum_{\substack{i \in [\text{In,Ga}] \\ j \in [\text{surf,bulk}]}} \mu_j^i b_j^i = 0 , \tag{6.2}$$

where $b_{\text{bulk}}^{\text{In}} = b_{\text{surf}}^{\text{Ga}} = -1$ and $b_{\text{surf}}^{\text{In}} = b_{\text{bulk}}^{\text{Ga}} = 1$ are the stoichiometric coefficients. Using the partition function Z of the system and $F = -kT \ln Z$, where k is the Boltzmann constant and T is the temperature, (6.1) and (6.2) result in a mass-action law

$$-\frac{\Delta F_0}{kT} = \sum_{\substack{i \in [\text{In,Ga}] \\ j \in [\text{surf,bulk}]}} b_j^i \ln N_j^i = \sum_{\substack{i \in [\text{In,Ga}] \\ j \in [\text{surf,bulk}]}} b_j^i \ln x_j^i \; , \tag{6.3}$$

where $x_j^i = N_j^i / \tilde{N}$ is the concentration of element i in layer j, $\tilde{N}$ being the number of available sites in the metal sublattice. ΔF_0 is the change of the free energy "per reaction", often called the "segregation energy" E_{S}. Using the condition that the metal sublattice is completely filled with In and Ga atoms by putting

$$x_j^{\text{Ga}} = 1 - x_j^{\text{In}} \; , \tag{6.4}$$

we obtain McLean's [9] equation [7],

$$\frac{x_{\text{surf}}^{\text{In}} \left(1 - x_{\text{bulk}}^{\text{In}}\right)}{x_{\text{bulk}}^{\text{In}} \left(1 - x_{\text{surf}}^{\text{In}}\right)} = \exp \left(-\frac{E_{\text{S}}}{kT}\right) \; . \tag{6.5}$$

A phenomenological description of segregation in InGaAs was given by Muraki et al. [10], based on experimental data obtained with secondary-ion mass spectroscopy (SIMS) and PL. Muraki et al. found a significant increase in segregation with increasing temperature and with decreasing As pressure. Note that the temperature dependence is opposite to that expected from (6.5); this shows the existence of a kinetic limitation on the segregation process [11, 12, 13]. Muraki et al. assumed, in their description of the segregation, that a certain fraction R of In atoms on the top surface layer (corresponding to the "bulk layer" in the description of Moison et al. [7]) segregate to the next layer (the "surface layer" in the description of Moison et al.). The quantity R is called the *segregation efficiency*. The remaining portion $(1 - R)$ of the In atoms is incorporated into the bulk before the next ML is completed. The In concentration in the nth layer is given by this model in the form

$$x_{\text{bulk}}^{\text{In}}(n) = \begin{cases} 0 & : \; n < 1 \\ x_0(1 - R^n) & : \; 1 \leq n \leq N \; , \\ x_0(1 - R^N)R^{n-N} & : \; n > N \end{cases} \tag{6.6}$$

where N is the amount of deposited In in units of ML of $\text{In}_{x_0}\text{Ga}_{1-x_0}\text{As}$, and x_0 is the nominal In concentration. Consequently, the In concentration in the surface layer (ML $n + 1$) is given by

$$x_{\text{surf}}^{\text{In}}(n + 1) = \frac{R}{1 - R} x_{\text{bulk}}^{\text{In}}(n) \; . \tag{6.7}$$

Examples of profiles that demonstrate the effect of segregation are given in

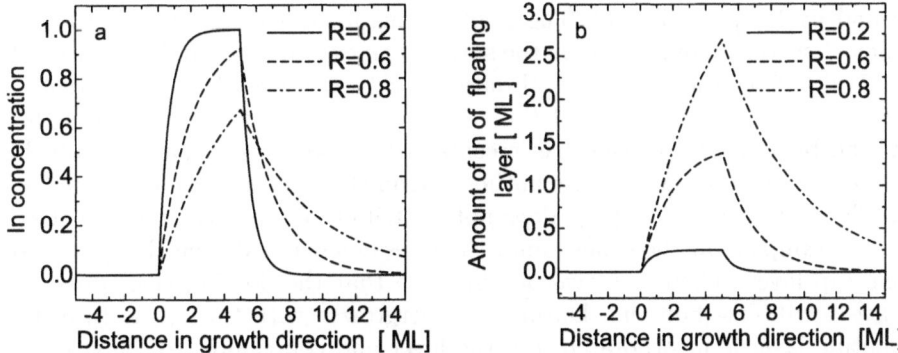

Fig. 6.2. (a) Examples of profiles of a (fictitious) $In_xGa_{1-x}As$ layer that demonstrate the effect of segregation. The curves were calculated from (6.6) using $N = 5$ ML, $x_0 = 1$ and $R = 0.2, 0.6$ and 0.8. (b) Amount of In in the floating layer on the surface, computed from (6.7)

Fig. 6.2. It is obvious from (6.7) that the In concentration in the surface layer can exceed 1 when $x_{bulk}^{In} > (1-R)/R$. For a typical value of $R = 0.85$ (see [10], growth temperature $T_G = 500°C$), an In concentration larger than 1 occurs for $x_{bulk}^{In} > 0.18$. In [10], it is assumed that this might be physically unrealistic. However, later experimental observations indeed revealed an amount of In of more than 1 ML (up to 1.8 ML) in the surface layer. These investigations started with a paper by Gerard [14], who found a method to probe the amount of In in the surface layer by using a shift of the 2D→3D transition that is usually observed at 1.8 ML in the growth of binary InAs. If a predeposition of InGaAs is used, the transition is shifted towards a smaller coverage because the In in the surface layer contributes to the growth of the InAs. In [15], this method was confirmed and the authors showed that it was possible to grow an abrupt InGaAs-on-GaAs interface by the predeposition of a 1 ML thick In-rich prelayer of InGaAs. Toyoshima et al. [16] used Gerard's method to probe x_{surf}^{In} and found that $0 \leq x_{surf}^{In} \lesssim 1.6$. The growth temperature was 520°C. If the amount of In exceeded ≈ 1.6 ML, a 2D→3D transition was observed for $x_0 = 0.27$ and 0.31. The concentration x_{surf}^{In} of In atoms on the surface during InGaAs growth as a function of the InGaAs thickness could be fitted well by (6.7) for $x_0 = 0.15, 0.2, 0.23$ and 0.25 ($R = 0.84$), as well as for $x_0 = 0.27$ and 0.31 with $R = 0.85$, for thicknesses below the 2D→3D transition. In a later paper, Toyoshima et al. showed [17], by variation of the growth temperature and As_4 pressure, that the amount of In atoms on the surface is crucial in determining when the transition in the growth mode takes place. For $x_0 = 0.34$, the 2D→3D transition was observed at 1.8 ML for a growth temperature of 520°C and an As_4 pressure of 1.3×10^{-5} Torr. The growth mode transition was prevented when the growth temperature was increased and/or the As_4 pressure was increased such that $\lim_{n \to \infty} x_{surf}^{In}(n) < 1.8$. An amount of In larger than 1 ML was also found by Kaspi et al. [18]. The publications mentioned above clearly show that, first, the amount of In on the surface can exceed 1 ML and that, second, (6.7) describes the experimental data for x^{In}_{surf} well even for $x^{In}_{surf} > 1$. As a consequence, the structure of the "surface layer" has to deviate from that of the bulk material. This has recently been confirmed experimentally by Garcia et al. [19], where those authors found that the "surface layer" (called a "floating-In-layer" in [19]) does not contribute to the accumulated stress in the sample. On the other hand, the stress increases during the growth of the cap layer. These observations indicate that the floating layer does not have a bulk-like bonding. During the overgrowth with GaAs, the In of the floating layer is incorporated into the layer and contributes to the stress in the sample.

In [13], a growth simulation of InGaAs/GaAs is presented that is based on a complex system of rate equations taking into account segregation. In accordance with the experimental results mentioned above, the floating layer is assumed to be a layer of physisorbed molecules in a weakly bound precursor state with van der Waals-type bonding. The process of segregation of bound In atoms into the physisorbed layer is modeled as an Arrhenius-type process with a frequency factor and an activation energy. The authors of [13] showed that the growth simulation results in segregation coefficients R that decrease with decreasing temperature, in agreement with the experimental data. Similar results were obtained earlier [12] by Monte Carlo simulations.

References

1. Y. Arakawa, H. Sakaki: Appl. Phys. Lett. **40**, 939 (1982)
2. M. Grundmann, O. Stier, D. Bimberg: Phys. Rev. B **52**, 11969 (1995)
3. A. Rosenauer, U. Fischer, D. Gerthsen, A. Förster: Appl. Phys. Lett. **71**, 3868 (1997)
4. D. Bimberg, M. Grundmann, N.N. Ledentsov: *Quantum Dot Heterostructures*, 1st edn. (Wiley, Chichester, 1999)
5. H.T. Dobbs, D.D. Vvedensky, A. Zangwill: Appl. Surf. Sci. **123/124**, 646 (1998)
6. S. Ruvimov, Z. Liliental-Weber, N.N. Ledentsov, M. Grundmann, D. Bimberg, V.M. Ustinov, A.Yu. Egorov, P.S. Kop'ev, Zh.I. Alferov, K. Scheerschmidt: Mater. Res. Soc. Symp. Proc. **421**, 133 (1996)
7. J.M. Moison, C. Guille, F. Houzay, F. Barthe, M. Van Rompay: Phys. Rev. B. **40**, 6149 (1989)
8. J. Nagle, J.P. Landesmann, M. Larive, C. Mottet, P. Bois: J. Cryst. Growth **127**, 550 (1993)
9. D. McLean: *Grain boundaries in metals*, 1st edn. (Clarendon Press, Oxford, 1957)
10. K. Muraki, S. Fukatsu, Y. Shirakia, R. Ito: Appl. Phys. Lett. **61**, 557 (1992)
11. O. Dehaese, X. Wallart, F. Mollot: Appl. Phys. Lett. **66**, 52 (1995)
12. N. Grandjean, J. Massies, M. Leroux: Phys. Rev. B **53**, 998 (1996)
13. G. Colayni, R. Vankat: J. Cryst. Growth **211**, 21 (2000)

14. J.-M. Gerard: Appl. Phys. Lett. **61**, 2096 (1992)
15. J.-M. Gerard, G. Le Roux: Appl. Phys. Lett. **62**, 3452 (1993)
16. H. Toyoshima, T. Niwa, J. Yamazaki, A. Okamoto: Appl. Phys. Lett. **63**, 821 (1993)
17. H. Toyoshima, T. Niwa, J. Yamazaki, A. Okamoto: J. Appl. Phys. **75**, 3908 (1994)
18. R. Kaspi, K.R. Evans: Appl. Phys. Lett. **67**, 819 (1995)
19. J.M. Garcia, J.P. Silveira, F. Briones: Appl. Phys. Lett. **77**, 409 (2000)

7 In$_{0.6}$Ga$_{0.4}$As/GaAs(001) SK Layers

In this chapter, we report a detailed structural and chemical study of buried and free-standing In$_{0.6}$Ga$_{0.4}$As SK islands [1]. The layers were grown by MBE on GaAs(001) substrates. We investigated two different types of samples, with nominal In$_{0.6}$Ga$_{0.4}$As layer thicknesses of 1.5 and 2 nm. The growth was interrupted for 0, 60 or 180 s prior to the deposition of a 10 nm thick GaAs cap layer. The chemical morphology of the buried layers was evaluated by the CELFA method. The free-standing islands were investigted by strain state analysis combined with FE calculations. The density and size distribution of the islands were obtained by conventional plan-view TEM.

We found two types of islands: coherent islands with a lateral size of approximately 13 nm, and large islands (40–100 nm) showing plastic strain relaxation. The density of the defect-free small islands decreases with increasing duration of the growth interruption, whereas the density and size of the large islands increase. A detailed study of the WL by the CELFA method revealed an In$_x$Ga$_{1-x}$As layer about 4 nm thick. The total amount of In contained in the WL decreases with increasing duration of the growth interruption. Composition profiles in the growth direction were measured. Their shape is explained mainly by three effects: segregation of In, incorporation of migrating In into the growing cap layer, and strain-driven migration of In and Ga. An inhomogeneous In concentration, increasing from bottom to top, is observed in free-standing islands.

7.1 Experimental Setup

The epitaxial growth of the heterostructures was performed by A. Förster of the Forschungszentrum Jülich. The samples were grown by MBE in a Varian Mod Gen II system on GaAs(001) substrates. The structures consisted of a GaAs buffer layer and an In$_x$Ga$_{1-x}$As layer with a nominal In concentration of $x = 60\%$. The overgrown samples contained a 10 nm thick GaAs cap layer. Two sets of samples, with nominal thicknesses of the In$_x$Ga$_{1-x}$As layer of 1.5 and 2 nm, were grown. For each set, three capped samples with growth interruptions of 0, 60 or 180 s after the deposition of the In$_x$Ga$_{1-x}$As were available. Additionally, an uncapped sample was grown, with an In$_x$Ga$_{1-x}$As layer thickness of 1.5 nm. The 0.1 μm thick GaAs buffer layer was deposited at a substrate temperature of 600°C, while the temperature was reduced to 500°C for the growth of InGaAs and the GaAs cap layer. The GaAs and InGaAs growth rates were 1 μm/h and 0.2 μm/h, respectively. The beam-equivalent pressure (BEP) V/III ratio was 22 for the GaAs growth, and 82 for the InGaAs growth.

TEM cross-section samples with {100} and {110} surfaces were prepared conventionally. In the final stage, Ar$^+$ or Xe$^+$ ion milling was used at an energy of 3 keV in a liquid-nitrogen-cooled specimen holder. Plan-view samples were prepared by dimple grinding and subsequent back-surface chemical etching in a solution of composition 1H$_2$O$_2$(30%):5NaOH. We used a Philips CM200 FEG/ST electron microscope with a spherical-aberration constant of $C_S = 1.2$ mm and a Scherzer resolution of 0.24 nm. The off-axis cross-section images used for the CELFA method were recorded with an on-line CCD camera with 1024×1024 picture elements. The specimen tilt was $(3 \pm 1)°$ away from a $\langle 010 \rangle$ ZA. The HRTEM images for the strain state analyses were exposed in a $\langle 110 \rangle$ ZA orientation on photographic negative film. The negatives were digitized with an off-line CCD camera with 1024×1024 picture elements.

PL investigations were carried out by R. Schmidt at the Forschungszentrum Jülich. They were performed at a temperature of 2.6 K with an excitation density of 4 mW/cm^2 ($\lambda_{Ex} = 517$ nm).

7.2 Experimental Results

7.2.1 Uncapped Samples

In this section we present the experimental results for the uncapped sample with an In$_x$Ga$_{1-x}$As layer thickness of 1.5 nm. Figure 7.1 depicts a plan-view weak-beam image. It shows two types of islands. We find coherent,

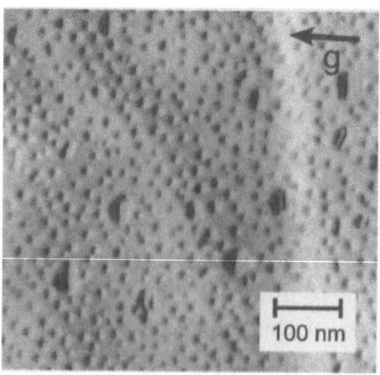

Fig. 7.1. Plan-view **g**/3**g** weak-beam image with **g** $= \langle 220 \rangle$, of the uncapped sample with an In$_x$Ga$_{1-x}$As layer thickness of 1.5 nm

small islands with a lateral size of 16 ± 3 nm, and large islands approximately 50 nm in size that contain misfit dislocations. The density of the coherent islands is 1.5×10^{11} cm^{-2}, and that of the large strain-relaxed islands is 2.7×10^{9} cm^{-2}.

HRTEM images revealed an average lateral size l of the coherent islands of (13.3 ± 1.5) nm and a height h of (2.8 ± 0.4) nm. The values were obtained from an evaluation of 15 islands. The errors were calculated from the standard deviation. The aspect ratio l/h of the islands is 4.8 ± 0.4.

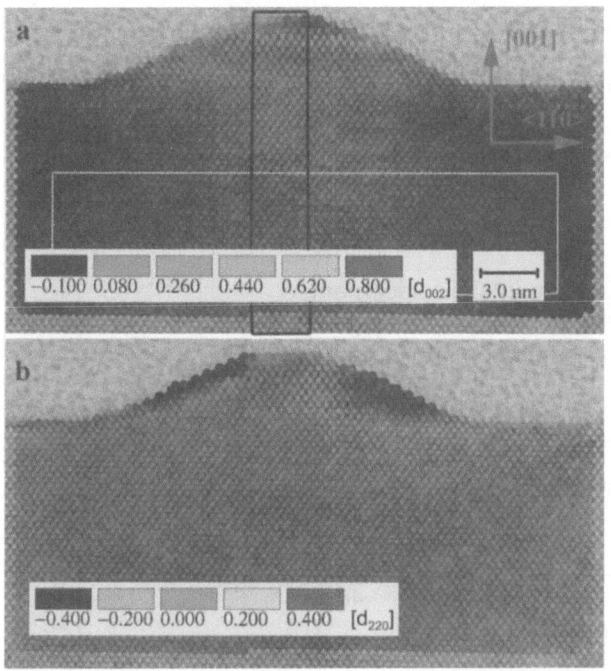

Fig. 7.2. (a) Color-coded map of local displacements in the growth direction evaluated from a $\langle 110 \rangle$ HRTEM image of an uncapped sample with an In$_x$Ga$_{1-x}$As layer thickness of 1.5 nm. The displacement values shown in the legend have been normalized with respect to the averaged spacing d_{002} of the horizontal (002) lattice planes inside the reference region, marked by a *white frame*. The *black frame* marks the area that was used to adapt FE-simulated and experimental displacements. (b) Components of the displacement vectors parallel to the interface, in the same area. The displacement values have been normalized with respect to the average spacing d_{220} of the vertical (220) lattice planes inside the reference region [Please see Plate 20 for color reproduction of this plate]

Figure 7.2 shows a DALI evaluation of an HRTEM image of a coherent island. The evaluated local displacement vectors have been decomposed into two perpendicular components. Figure 7.2a displays a color-coded map of the component in the growth direction. Figure 7.2a clearly shows that the displacements (their mean value vanishes inside the reference region) increase from the bottom to the top of the island, revealing an enlarged lattice parameter compared with the GaAs buffer. The black frame marks the area that was used to average the local displacements along the horizontal (002) planes, yielding the displacement profile in the growth direction shown in Fig. 7.3 (solid circles). Note that the displacements near the surface beside the island do not provide a strong indication of a WL. Figure 7.2b shows the components of the local displacement vectors parallel to the interface. The red region corresponds to displacement vectors pointing to the right, and the blue region corresponds to those pointing to the left. Both regions result from the relaxation of the elastic strain of the island, which results in a displacement of atoms near the island's surface in the outward direction.

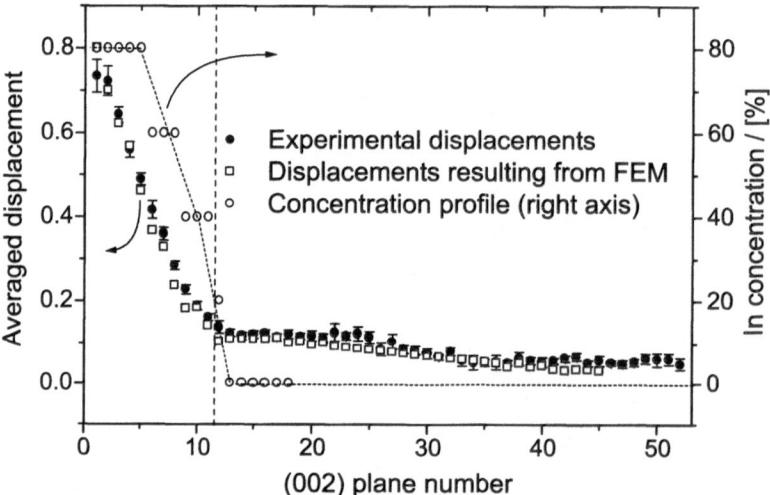

Fig. 7.3. Experimentally measured and FE-simulated average displacements plotted versus the (002) plane number. The displacements were averaged over a region corresponding to the black frame in Fig. 7.2a. The *open circles* show the concentration profile that was used for the FE calculation. The *vertical dashed line* indicates the position of the surface in the vicinity of the island

The FE model that was generated in accordance with the island shape and the local sample thickness measured by the QUANTITEM [2, 3, 4] procedure (see Sect. 4.2) was shown earlier in Fig. 4.17. It should be mentioned that the local specimen thickness could be evaluated in the GaAs buffer only, because the HRTEM contrast pattern depends on the In concentration also. Figure 7.4 is a plot of the specimen thickness, which was used for the FE modeling, revealing a wedge-shaped specimen. The angle of the wedge is 26°. This specimen shape is in a good agreement with that expected from the

TEM sample preparation conditions, where the incidence angle of the Ar$^+$ ions was 15° from both sides, so that we expect a 30° wedge. As shown in Fig. 7.4, the specimen thickness in the island region was extrapolated from the values measured in the GaAs buffer.

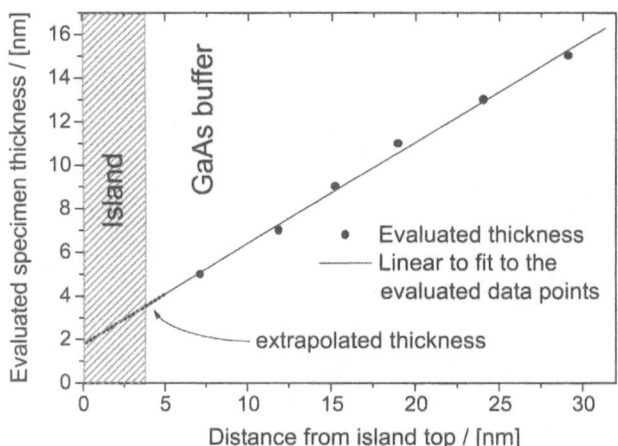

Fig. 7.4. Evaluated specimen thickness plotted versus the distance from the top of the island, parallel to the growth direction. The specimen thickness in the region of the island was extrapolated from the GaAs buffer

Figure 7.3 contains the concentration profile that leads to the best agreement between the measured and simulated averaged displacements, which are also shown in Fig. 7.3. The concentration profile shows four steps, because the FE model of the island and of the WL was subdivided into five solids. The In concentration is homogeneous inside each solid. The profile shows only four steps, because the two solids next to the top of the island have the same In concentration. From the position of the surface beside the island, we estimate that the thickness of the WL is roughly 1 ML (see vertical dashed line in Fig. 7.3).

To be able to compare the strain fields of the whole FE model with the experimental results, we evaluated the 2D model of the projected atom columns (see Sect. 4.3.2) analogously to the experimental image. The result was shown earlier in Fig. 4.19. Obviously, the evaluated In concentration profile leads to good agreement with the experimental displacement vectors shown in Fig. 7.2.

7.2.2 Capped Samples

7.2.2.1 Structural Properties

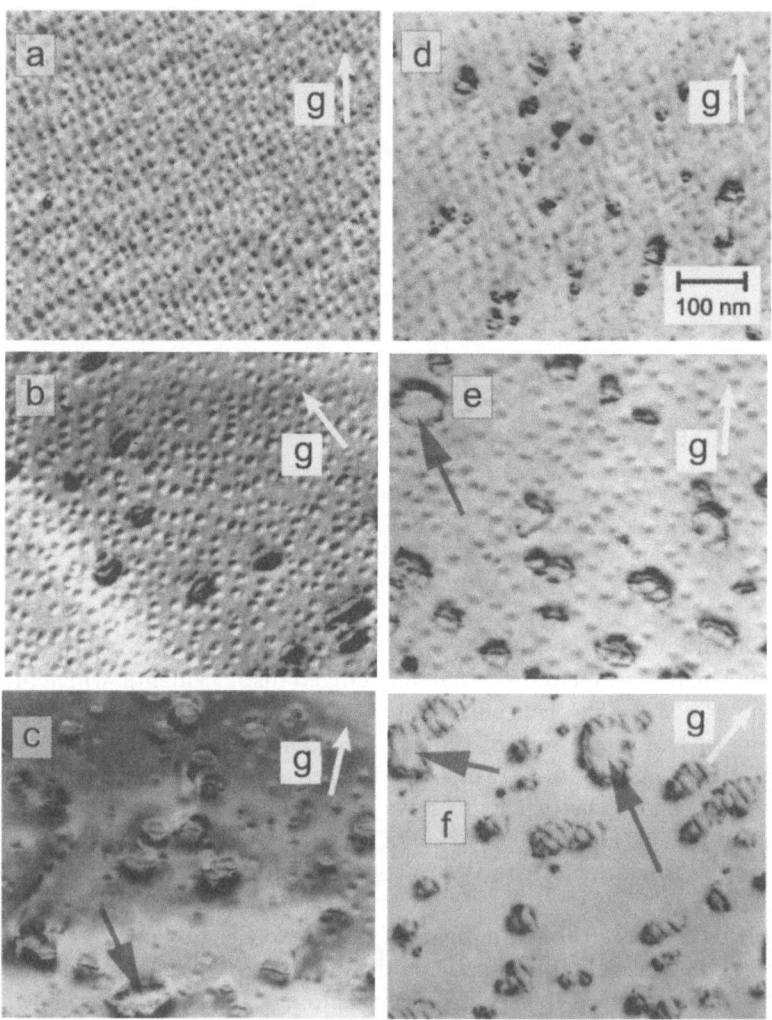

Fig. 7.5. TEM plan-view micrographs from capped samples obtained under a $g/3g$ weak-beam condition with $g = \langle 220 \rangle$. Micrographs (**a**), (**b**) and (**c**) correspond to an $In_xGa_{1-x}As$ layer thickness of 1.5 nm and (**d**), (**e**) and (**f**) to a thickness of 2 nm. The duration of the growth interruption was 0 s for (**a**) and (**d**), 60 s for (**b**) and (**e**), and 180 s for (**c**) and (**f**). The *dark arrows* mark islands with missing strain or dislocation contrast in the center

(a) TEM Plan-View Images. Figure 7.5 shows weak-beam plan-view micrographs of all samples investigated that contain a 10 nm thick GaAs cap layer. Similarly to the uncapped sample, we find two types of islands: coherent islands with a size of approximately 15 nm, and strain-relaxed islands that reach a lateral size of about 100 nm for the longest duration of the growth interruption. It is obvious from Fig. 7.5 that the small, coherent islands are not stable. For both of the $In_xGa_{1-x}As$ layer thicknesses of 1.5 and 2 nm, the density of the coherent islands decreases with increasing duration of the growth interruption. The density of the large, strain-relaxed islands depends inversely on the duration of the interruption. We did not observe any coherent islands in the 2 nm sample with a growth interruption of 180 s.

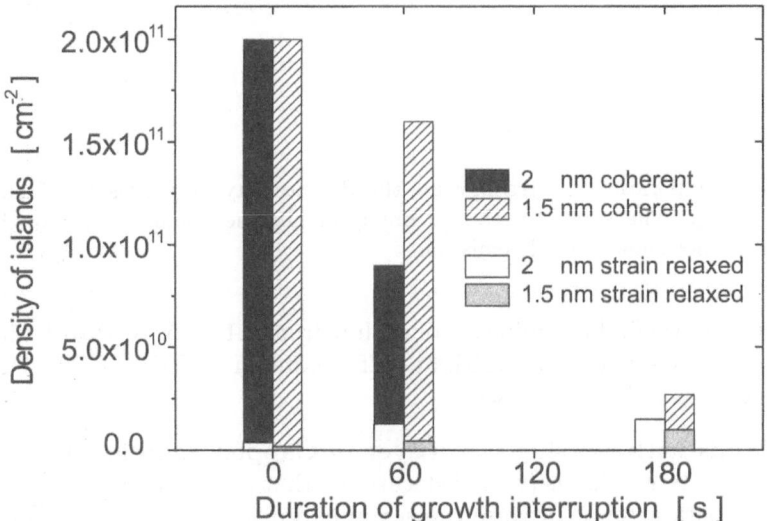

Fig. 7.6. Density of capped islands investigated, plotted versus the duration of the growth interruption

Figure 7.6 gives a survey of the island densities. It reveals that the densities of relaxed islands are significantly larger for the 2 nm samples. Furthermore, one can clearly recognize that the density of the coherent islands drops more quickly in the 2 nm samples. In this context, it is important to note that the initial density of coherent islands was equal for the two $In_xGa_{1-x}As$ layer thicknesses. Therefore, we deduce that the initial coherent islands are more stable in the 1.5 nm sample than in the 2 nm sample. This result will be important later on.

(b) PL. The low stability of the islands in the 2 nm sample is also visible in the PL spectra shown in Fig. 7.7. Here we find that the position of the QD emission line is approximately stable at 1.173 eV in the 1.5 nm sample. In

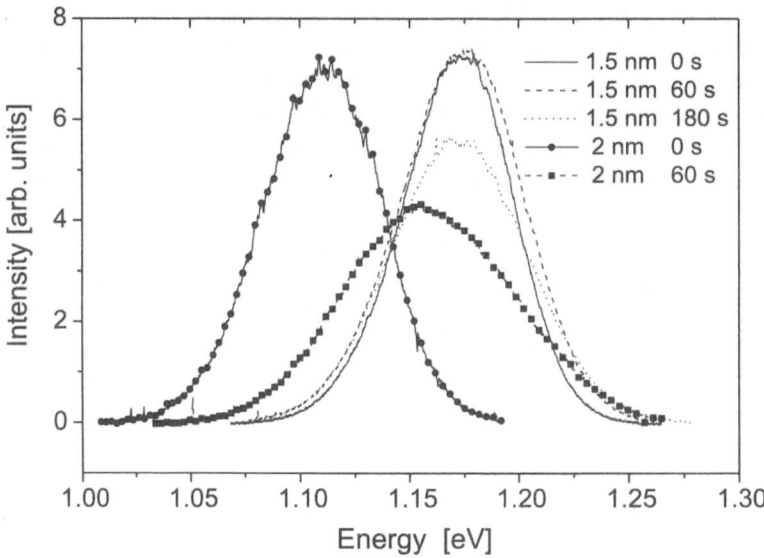

Fig. 7.7. Low-temperature spectra showing the PL intensity from QDs as a function of the energy. The sample with a 2 nm layer thickness and 180 s growth interruption did not show any QD emission

the 2 nm sample, we find a significant blue shift from 1.113 eV (0 s) to 1.155 eV (60 s). Additionally, the full width at half maximum (FWHM) of the PL peak of the 60 s sample is increased.

7.2.2.2 Islands with Missing Cap Layer in Capped Samples. In Fig. 7.5, we find some large islands (with dislocations) that are rather conspicuous (marked with dark arrows) because the strain and dislocation contrast vanishes in an approximately circular area around the island center. Such islands occur mainly in the 2 nm samples with 60 and 180 s growth interruptions. In corresponding HRTEM images, we find islands with missing cap layers on their top at a similar density. One of these images is displayed in Fig. 7.8a. The imaged crystal region contains two misfit dislocations with terminating {111} lattice planes of the substrate.

Figure 7.8b shows the components $u_\perp$ of the displacement vectors in the growth direction evaluated from the HRTEM image in Fig. 7.8a (inside the region indicated by a black frame). The sharp transition from green ($u_\perp \approx +0.5$) to blue ($u_\perp \approx -0.5$) along a horizontal line in the right part of the image is due to a missing (002) lattice plane terminating at a misfit dislocation. In the left part of the image, the increasing value of $u_\perp$ clearly supplies evidence for the existence of an In$_x$Ga$_{1-x}$As island extending from the interface to the top of the visible contrast pattern in Fig. 7.8a. Hence, the GaAs layer is completely missing at the top of the island. A profile of $u_\perp$ along the growth direction, averaged inside the narrow region marked by a

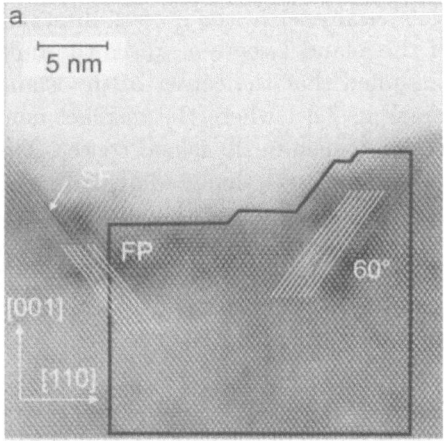

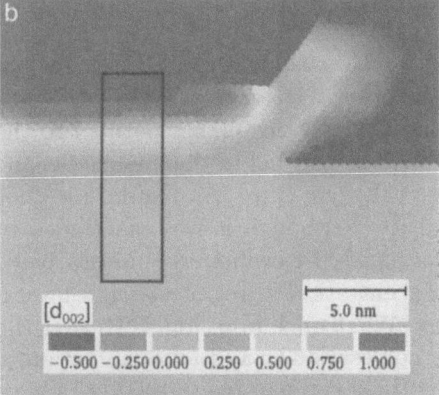

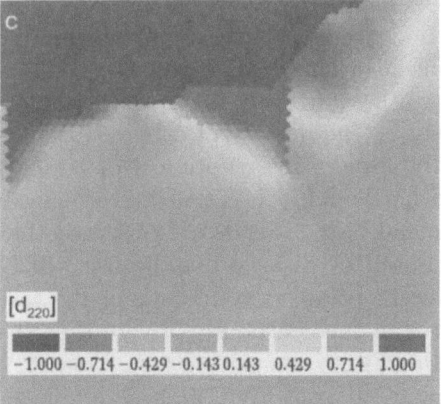

Fig. 7.8. (a) HRTEM micrograph of a sample with a 2 nm $In_xGa_{1-x}As$ layer thickness taken in the $[\bar{1}10]$ ZA orientation, showing an island with a missing cap layer. The island contains a Frank partial (FP) dislocation and a 60° dislocation. The *red lines* help to identify terminating {111} lattice planes of the substrate. The *black frame* indicates the region that was evaluated by strain state analysis. The color-coded maps (**b**) and (**c**) show local displacement vector components in the growth direction and in the interface direction, respectively. The reference region was chosen to be inside the GaAs buffer. In (**c**), the abrupt transitions from *green* to *blue* and from *red* to *green* along two vertical lines at the *left* and *right* sides, respectively, of the island are due to (220) lattice fringes of the substrate that end at the dislocation cores. The *red region* corresponds to displacement vectors pointing to the *right*, and the *blue* region to displacement vectors pointing to the *left* [Please see Plate 21 for color reproduction of this plate]

black frame in Fig. 7.8b, reveals a lattice parameter in the growth direction of $a_\perp = 1.094 a_{GaAs}$ close to the top of the island (where a_{GaAs} is the bulk lattice parameter of GaAs). We have assumed that the center of the island is fully strained, as can be verified from Fig. 7.8c, where the displacement vector components in the interface direction vanish in the island center. The corresponding bulk material lattice parameter a was calculated according to (4.26), where the parameter α_R depends on the degree of the elastic relaxation in the electron beam direction. Assuming that the top of the island is completely strain-relaxed in the electron beam direction, we have obtained [5] $\alpha_R \approx 1.5$, yielding $(a - a_{GaAs})/a_{GaAs} \approx 0.063$, which corresponds to an In concentration of approximately 90%.

7.2.2.3 Evaluation of the Chemical Composition In this section, we present the evaluation of the chemical composition of the capped samples by exploitation of the chemically sensitive {020} reflections.

(a) Conventional DF Imaging. Figure 7.9 shows single-beam DF images obtained with a strongly excited (002) reflection. Note that this corresponds to diffraction of the electron beam by the (002) lattice planes parallel to the interface plane. Therefore, the electron beam is parallel to the interface plane to a very good approximation and the visible In$_x$Ga$_{1-x}$As region does not contain any blurring induced by crystal tilt. Although it was mentioned in Sect. 5.1.2.2 that images of this kind can hardly be evaluated quantitatively, they provide a good survey of the In concentration in regions with a lateral extent of several hundreds of nanometers. We deduce from Fig. 5.1b that the darkest regions of the In$_x$Ga$_{1-x}$As layer correspond to an In concentration of approximately 22%. Regions with a brightness comparable to the GaAs have an In concentration of 40%. Therefore, the dark stripe with the bright dots corresponds to a quantum well with an In concentration below approximately 30%, containing islands with an In concentration of approximately 40% or larger. Note that the concentrations given here are only rough estimates.

As expected from the plan-view images, the density of coherent islands decreases with increasing duration of the growth interruption. Not a single coherent island is found in the sample with a 2 nm In$_x$Ga$_{1-x}$As layer thickness and a growth interruption of 180 s. Instead, large islands are observed that frequently exhibit a tangled contrast caused by dislocations. Figure 7.9f (2 nm layer thickness and 180 s growth interruption) contains one of the rare islands that appear to be dislocation-free in the central part. Note that the cap layer is interrupted on top of the island.

The most striking result that is apparent from Fig. 7.9 is a significant transformation of the In$_x$Ga$_{1-x}$As layer morphology by the GaAs overgrowth. The initially 1 or 2 ML thick WL is significantly broadened so that the islands and WL now have similar dimensions in the growth direction. Therefore, the initial In$_x$Ga$_{1-x}$As layer consisting of a thin WL and 3D islands has been

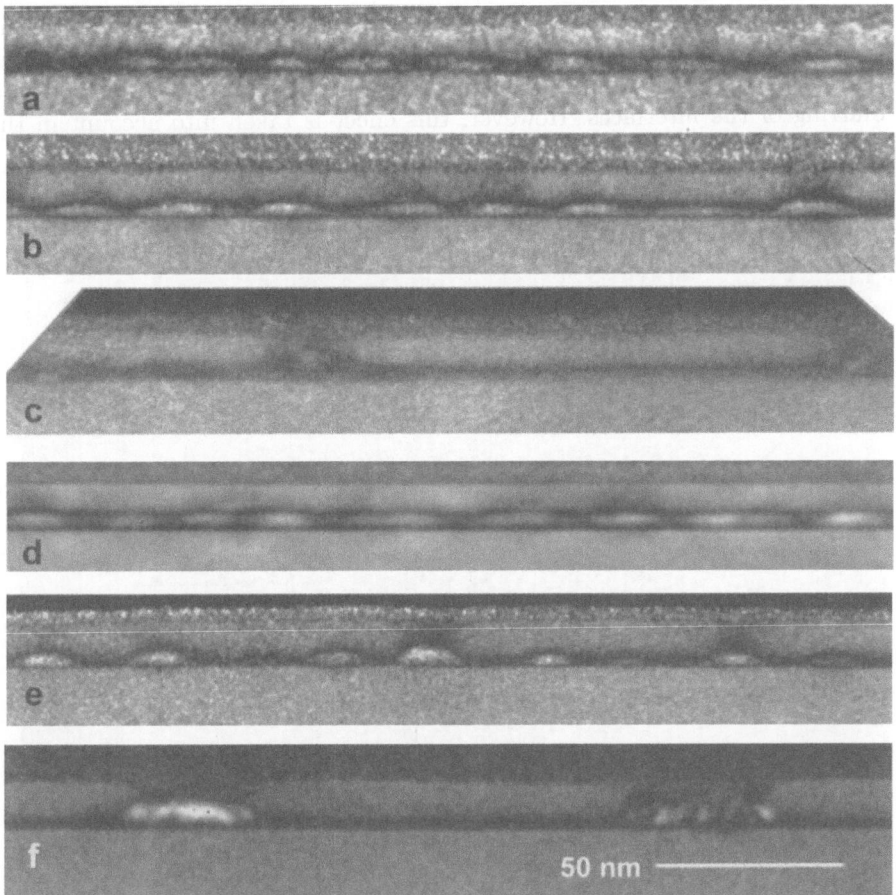

Fig. 7.9. Conventional single-beam DF images obtained with a strongly excited (002) beam on the optic axis. (**a**), (**b**) and (**c**) show the results for the sample with 1.5 nm $In_x Ga_{1-x}As$ layer thickness and (**d**), (**e**) and (**f**) those for the 2 nm sample. The duration of the growth interruption was 0 s for (**a**) and (**d**), 60 s for (**b**) and (**e**), and 180 s for (**c**) and (**f**). The corresponding In concentration can be roughly estimated from Fig. 5.1b

transformed into an approximately 4 nm thick quantum well with a low In concentration that contains inclusions with higher concentrations.

(b) CELFA. An investigation of the local composition of the capped layers was performed by the CELFA method. In order to avoid effects of variations of the lattice parameter on the observed contrast pattern, the samples were oriented in such a way that the (020) lattice fringes perpendicular to the interface plane could be used for the evaluation. These lattice fringes have the same lattice parameter, to a good approximation, throughout the regions where the $In_x Ga_{1-x}As$ layer had grown pseudomorphically. For this inveti-

gation, the samples were tilted by approximately 3° around an axis running parallel to the interface plane and perpendicular to the electron beam direction. Note that this sample orientation induces a small but not significant blurring of the interfaces. However, this effect is taken into account in the quantitative evaluations presented in the discussion.

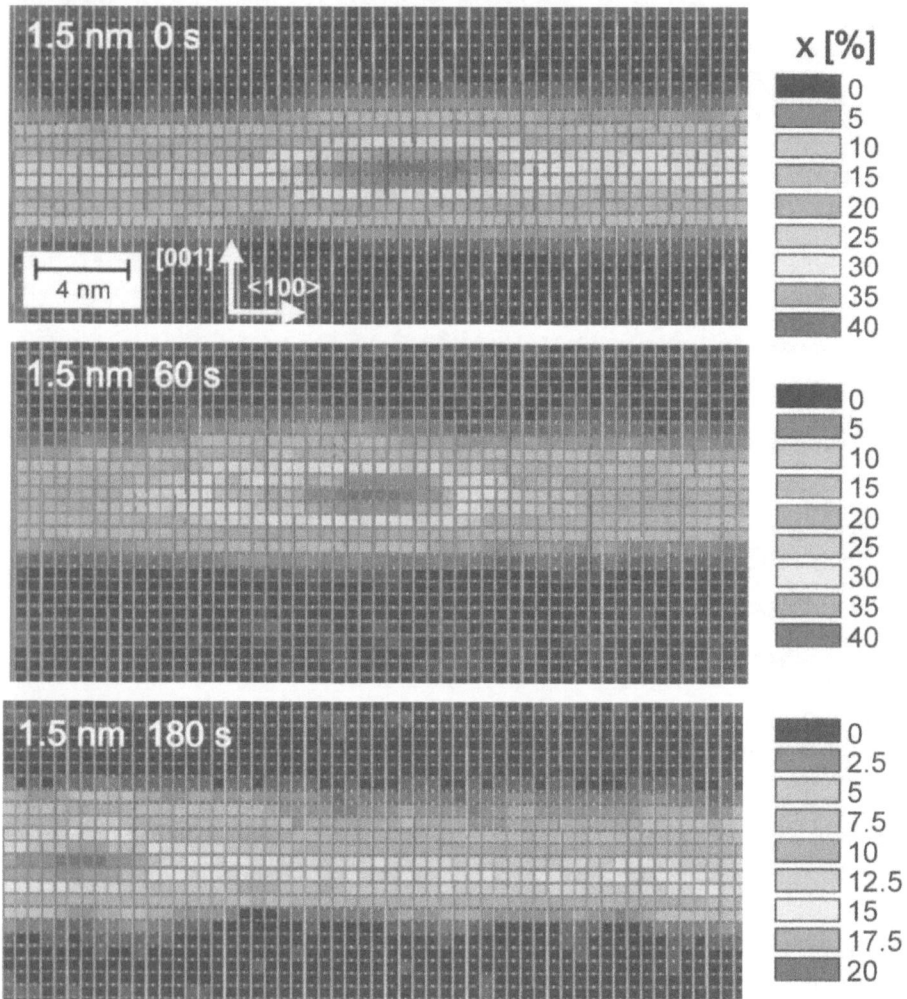

Fig. 7.10. Color-coded maps of the local In concentration x evaluated by the CELFA method. Note that the *red* color corresponds to an In concentration of 20% for the 180 s specimen (*bottom image*), whereas it corresponds 40% for the *upper two images* [Please see Plate 22 for color reproduction of this plate]

Figure 7.10 shows the resulting color-coded maps of the local In concentration. Each colored square covers an area of $a_{GaAs} \times a_{GaAs}$. This figure clearly reveals the existance of a broad WL with a maximum In concentration x that decreases with increasing duration of the growth interruption. The WL contains inclusions with an enlarged In concentration, with a lateral size of approximately 13 nm. For the samples with 0 and 60 s growth interruptions, the maximum measured In concentration is 40% in the 1.5 nm samples and 48% in the 2 nm samples. Small, coherent islands are also found in the 1.5 nm sample with a 180 s growth interruption. These islands show a maximum In concentration of approximately 18%. Note that the In concentration that is measured in the island regions is smaller than the real In concentration inside the buried island if the sample thickness in the electron beam direction is larger than the size of the island.

Figure 7.11 shows concentration profiles in the growth direction obtained from the WLs of all samples investigated. One can clearly recognize that the maximum In concentration decreases with increasing duration of the growth interruption. The profiles are not symmetrical, but show a slower decay towards the GaAs cap layer. This is a clear indication of segregation. The area under each curve yields the total amount of In that is contained in the WL. Figure 7.12 illustrates the behavior of this total amount of In as a function of the duration of the growth interruption. For both sets of samples, with 1.5 and 2 nm layer thicknesses, the amount of In contained in the WL can be described by an exponential decay. The time constant τ (see caption of Fig. 7.12) for the 1.5 nm samples is 1.5 times larger than that for the 2 nm samples. The extrapolation of the exponential fit curves towards longer growth interruptions approaches an asymptotic value corresponding to 2.2 ML $In_{0.6}Ga_{0.4}As$.

7.3 Discussion and Conclusions

In the previous section we found that the "wetting layer" between the islands of capped samples differs significantly from the WL that was observed in the uncapped sample. In the latter case we found 3D islands with a height of approximately 11 ML (see Fig. 7.3) measured from the WL surface. Strain state analysis of an uncapped island revealed indications of an approximately 1 or 2 ML thick WL. On the other hand, the investigations of the capped islands unambiguously showed an approximately 15 ML thick "wetting layer" that contains islands in the form of In-rich insertions. Here we deduce a growth model to explain the observed morphological transformation of the WL during the overgrowth with GaAs.

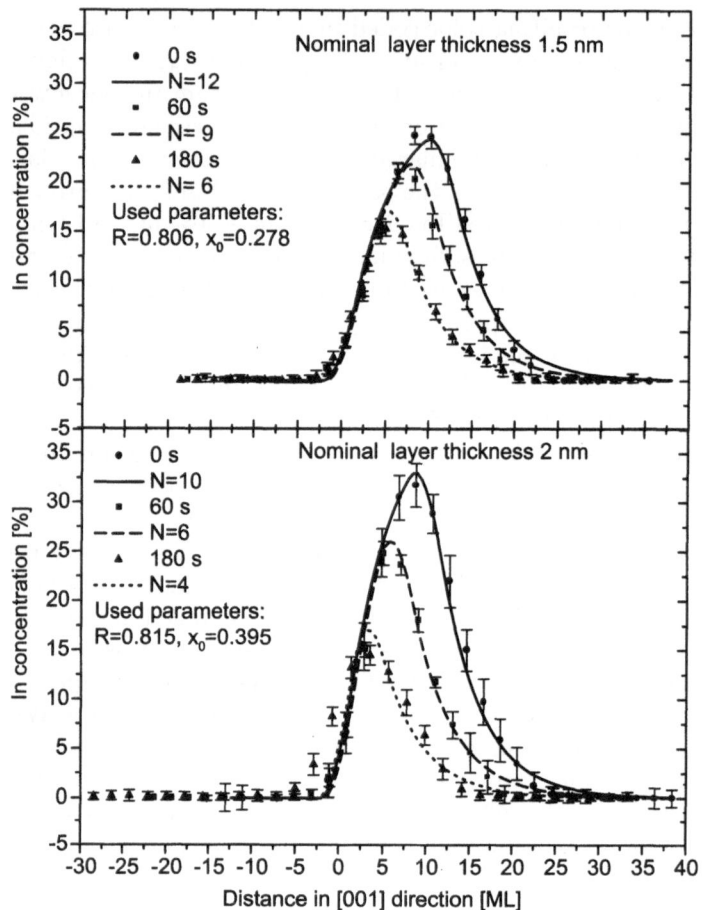

Fig. 7.11. Concentration profiles in the growth direction plotted versus the distance in the [001] direction. The profiles were obtained by averaging along the (002) planes in regions *of the WL* approximately 3 nm wide . The *dots, squares* and *triangles* represent the experimental data. The error bars were calculated from the mean deviation of the averaged values. The zero of the abscissa corresponds to the interface between the GaAs buffer and the In$_x$Ga$_{1-x}$As layer. The *solid, dashed* and *dotted curves* are fit curves calculated from the phenomenological Muraki formula for segregation. The meaning of the fit parameters N, R and x_0 is explained in the text

7.3.1 Strain-Induced Migration of Ga

In Figs. 7.9 and 7.10, it is conspicuous that the upper interface of the In$_x$Ga$_{1-x}$As layer appears flat. Therefore, we have to explain why the incorporation of migrating In takes place only between the islands and not on top of them. It can be seen in Fig. 7.8 that the cap layer does not grow on top of strain-relaxed islands. This effect was also observed by Xie et al. [6],

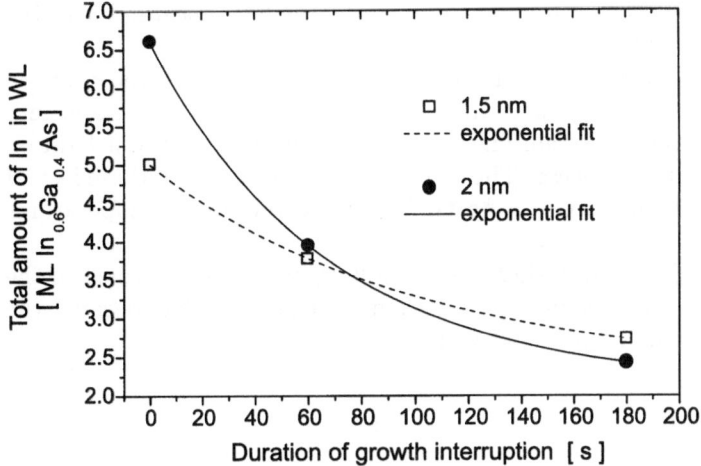

Fig. 7.12. Total amount of In X_{sum} contained in the WL, plotted versus the duration of the growth interruption t. The *solid* and *dashed lines* represent exponential fit curves calculated from the equation $X_{\text{sum}}[\text{ML In}_{0.6}\text{Ga}_{0.4}\text{As}] = X_0 + C \exp(-t/\tau)$. The fit parameters are $X_0 = 2.27$, $C = 2.73$, $\tau = 101.6$ for the 1.5 nm samples and $X_0 = 2.13$, $C = 4.47$, $\tau = 66.8$ for the 2 nm sample

where those authors found an island-induced strain-driven adatom migration during the GaAs cap layer growth, by the placement of very thin AlGaAs marker layers. Xie et al. found that the growth rate of the GaAs cap layer depends on the local in-plane lattice parameter at the growth surface and on the growth temperature. The elastic relaxation of coherent islands or the plastic relaxation of incoherent islands yields an enlarged in-plane lattice parameter on top of the islands. If the surface mobility of the Ga atoms is large enough ($T_{\text{G}} \geq 480°\text{C}$), the Ga atoms migrate from the top of the islands toward the regions between the islands. Therefore, the growth rate of the cap layer is significantly larger in between the islands.

7.3.2 Bulk Interdiffusion

Generally, the interdiffusion of In in GaAs could be expected to lead to a broadening of the In$_x$Ga$_{1-x}$As layer. The diffusion coefficient is $D = 1.6 \times 10^{-24}\text{cm}^2/\text{s}$ for the growth temperature of 500°C [7]. To test whether interdiffusion could contribute to the broadening of the wetting layer, we assumed a Heaviside function for the initial concentration profile $x(t, z)$, where t is the duration of the diffusion process and z is the coordinate in the growth direction. By calculating a solution of the linear diffusion equation

$$\frac{\partial x(t, z)}{\partial t} = D \frac{\partial^2 x(t, z)}{\partial z^2} , \tag{7.1}$$

we found that the effect of interdiffusion is negligible here because t is of the order of only a few minutes.

Additionally, it is conceivable that the strain has an effect on the diffusion in strained-layer heterostructures. In [8] the effect of strain was taken into account by regarding the strain energy as a contribution to the activation energy of the diffusion process. The authors of [8] found that the effect of strain is negligible in an In$_x$Ga$_{1-x}$As/GaAs heterostructure at temperatures below 600°C.

Note that interdiffusion during the ion milling process (performed in the final stage of the TEM specimen preparation) is also negligible, because the specimen heating is well below 300°C [9].

7.3.3 Segregation

To be able to compare the values found in the literature with our measurement, we fitted the experimentally observed concentration profiles in the growth direction (depicted in Fig. 7.11) with the segregation profile of (6.6). The parameters x_0, R and N were used as fit parameters. A tilt of the specimen of 4° relative to the exact ZA orientation was taken into account, and a specimen thickness of 15 nm was assumed. Figure 7.11 also contains the resulting fit curves and the corresponding values of the fit parameters. Note that the ascending part of the concentration profile mainly defines x_0, whereas the descending part determines R. In agreement with the data found in the literature, all curves are fitted well by $R = 0.810 \pm 0.006$. However, the fit values of $x_0 = 0.278$ for the 1.5 nm samples and 0.395 for the 2 nm samples deviate significantly from the nominal value $x_0 = 0.6$. Note that x_0 has been kept constant for samples with the same In$_x$Ga$_{1-x}$As layer thickness. This was found to be a good approximation for all measured profiles of the specimens with 0 and 60 s growth interruptions. In the samples with a 180 s interruption, a tendency to smaller values $x_0 \approx 0.2$ was observed.

In (6.6), the values of N describes the nominal In$_x$Ga$_{1-x}$As layer thickness, i.e. the number of ML that were grown under In flux. The fitted values of N are approximately 12, 9 and 6 for the 1.5 nm samples (in order of increasing duration of the growth interruption), and 10, 6 and 4 for the 2 nm samples. As was mentioned in Sect. 6.2, the In content of the descending part of the concentration profile gives the In content in the floating layer. The floating layer has a van der Waals-type bonding. It can contain an amount of In significantly larger than 1 ML. Hence, Fig. 7.11 reveals an amount of In in the floating layer that clearly decreases with increasing duration of the growth interruption. Section 7.2.2.1 stated that the density of coherent islands drops with increasing duration of the growth interruption. Therefore, both the floating layer and the dissolution of small, coherent islands feed large, plastically strain-relaxed islands whose density and average size increase during the growth interruption. Dissolving coherent islands can therefore be regarded as a source of In atoms that migrate over the surface towards the growing large islands, which act as sinks for In atoms.

7.3.4 Growth Model

On the basis of the results presented in the previous paragraphs, we have developed the following growth model, which leads to the observed morphology of the capped samples; this deviates significantly from that of the uncapped samples.

During the initial growth, the In concentration profile is determined by segregation until the 2D→3D transition is reached and the formation of 3D islands starts. Afterwards, it can be expected that further deposited In is aggregated in the islands. According to data in the literature, the 2D→3D transition occurs as soon as the amount of In in the floating layer reaches a value between 1.3 and 1.6 ML. The first of those values was obtained from studies of the growth of binary InAs, where the 2D→3D transition is observed after the deposition of 1.8 ML InAs [10], and the second value was obtained from studies of the epitaxy of $In_{0.31}Ga_{0.69}As$ [11]. For the growth of $In_{0.6}Ga_{0.4}As$, this corresponds to a total deposited amount of $In_{0.6}Ga_{0.4}As$ of (4.1 ± 0.6) ML.

Figure 7.12 shows that the amount of In contained in the WL is larger for the samples grown without interruption than for the samples with a growth interruption of 60 s, and is smallest after a growth interruption of 180 s. This result can be interpreted as follows. The dissolution of the small, coherent islands provides a current of In atoms that is incorporated during the initial stage of the cap layer growth (see Figs. 7.13a,b). This current decreases with increasing duration of the growth interruption, owing to the shrinking density of coherent islands. Starting with the same island density in samples without growth interruption, the densities decrease more quickly in the 2 nm samples. Therefore, the current of migrating In is stronger in the 2 nm samples than in the 1.5 nm samples. The dissolving islands may thus be regarded as a source of In atoms that is active even if the In shutter of the MBE apparatus has been closed. This could explain the large amount of In in the "WL" of the samples without growth interruption (i.e. the large values obtained for the fit parameter N) and the slope of the ascending parts of the segregation profiles that leads to fitted values of x_0 significantly smaller than $x_0 = 0.6$. It could also explain why the value $x_0 = 0.396$ found for the 2 nm samples is larger than the $x_0 = 0.278$ observed in the 1.5 nm samples, because the current of In atoms from dissolving islands is larger in the 2 nm samples. Note that the dissolving islands act as a source of In atoms until they are completely covered by the cap layer or are completely dissolved. This would suggest a correlation between the height of the islands and the height of the ascending parts of the segregation profiles, which would explain the value $N \approx 11$ ML obtained for the samples without growth interruption; this value is in a good

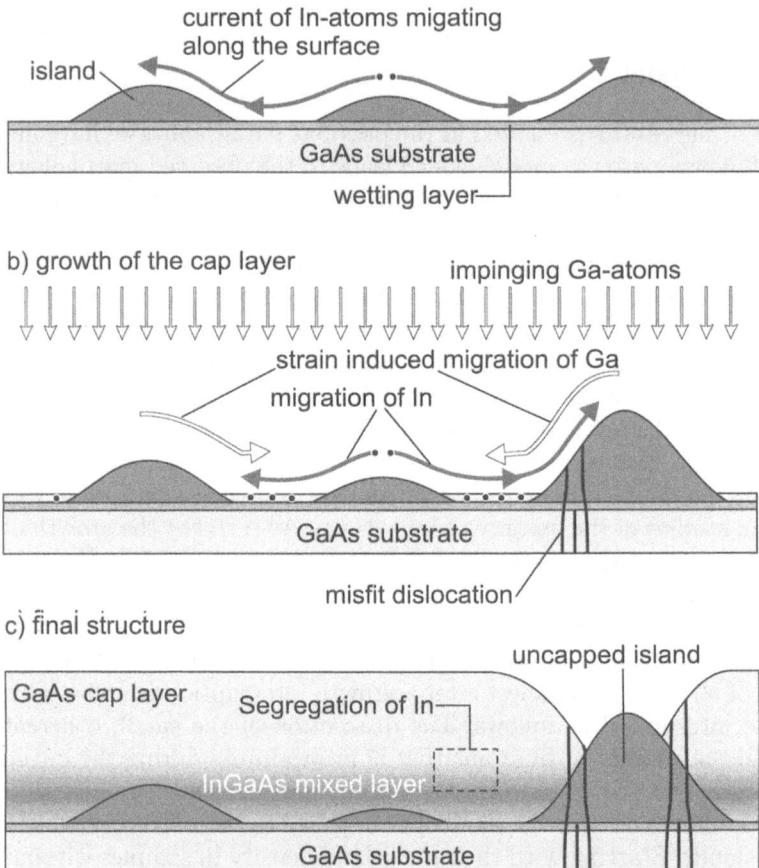

a) growth interruption

b) growth of the cap layer

c) final structure

Fig. 7.13. Schematic drawing showing a survey of the effects that influence the structure of the In$_x$Ga$_{1-x}$As SK layer. (**a**) Migration of In atom takes place during the growth interruption. The density of small, coherent islands decreases and the density and volume of large, strain-relaxed islands increase. (**b**) During the growth of the cap layer, the migration of In atoms still prevails, acting as a source of In atoms, which are incorporated into the growing cap layer. Additionally, strain-induced migration of Ga atoms occurs, leading to a decreased growth rate of the cap layer on top of elastically and plastically strain-relaxed islands. The In migration stops as soon as those islands which acted as a source of In atoms are covered. Subsequently, the In concentration decreases exponentially, owing to the effect of segregation. (**c**) Final layer structure, showing a broad In$_x$Ga$_{1-x}$As mixed layer with an increased amount of In atoms compared with the original WL caused by In atoms from dissolved coherent islands. Additionally, large plastically strain-relaxed islands occur with a missing cap layer. Undissolved coherent islands are visible as inclusions with an increased In concentration

agreement with the mean height of 10 ML of the coherent islands observed in uncapped samples (see Sect. 7.2.1).

During the growth interruption, In atoms from both the dissolving islands and the floating In layer contribute to the formation of large, plastically strain-relaxed islands (Fig. 7.13b).

Finally, the In concentration drops exponentially when the (remaining) floating In layer is incorporated into the cap layer, forming the descending part of the segregation profile (Fig. 7.13c), from which the segregation efficiency R can be obtained. Note that the ascending part of the profile determines x_0, which, according to the preceding discussion, cannot be interpreted as the nominally deposited concentration of In. Instead, x_0 is influenced by processes that are entangled in complex way, such as incorporation of In (and Ga) from dissolving islands, strain-induced migration of Ga from the tops of islands where the In concentration is larger than 0.6, and segregation, which correlates the In in the floating In layer with that in the bulk layer.

7.3.5 Correlation with PL Results

In Fig. 7.7, the large FWHM for the sample with a 2 nm $In_xGa_{1-x}As$ layer thickness and a growth interruption of 60 s is conspicuous. From this observation, we could deduce a broad variation of the island sizes and/or the In concentration inside the islands. Indeed, this expectation is confirmed in Fig. 7.9e, where one can see small islands with a low In concentration and larger islands with a high In concentration. In accordance with the PL data, Fig. 7.9e exhibits the largest differences in the sizes and In concentrations of the islands.

7.3.6 Composition Distribution in Free-Standing Islands

Theoretical considerations of the SK growth of islands during alloy deposition carried out by Tersoff [12] suggest that the islands nucleate at a substantially different composition than that of the alloy layer. Note that this statement refers to the critical nucleus, which is generally much smaller than the final island. During island growth, the WL constitutes a reservoir of In atoms that feeds the islands. Tersoff pointed out that the growth of the islands takes place at the expense of the film if the incident flux of In atoms is turned off. The compositional enrichment of the islands leaves behind a compositionally depleted film. This consideration could explain our strain state analysis measurements, where we could not find indications of a WL between the islands in uncapped samples. The bottom ML of the island, containing 20% In (see Fig. 7.3) could reflect the WL during the earliest stage of the growth, could have been "frozen" into the island. Tersoff discussed a possible "self-capping" of QDs that would result in islands with a high In concentration in the center, surrounded by material with a lower concentration. In our experiments, we did not find indications of "self-capping". Instead, the In concentration increases from the bottom to the top of the island (see Fig. 7.3). We suppose that the segregation of In and the strain-induced migration [6] of In and Ga, which was discussed above, is the main effect that defines the composition distribution inside an island. During the initial growth of an island (the first ML), the in-plane lattice parameter of the island adapts itself to the lattice parameter of the substrate. As growth of the island proceeds, the degree of elastic relaxation and, therefore, the in-plane lattice parameter increase. Owing to the strain-induced migration of In and Ga, the local composition of the currently growing ML M of the island depends on the in-plane lattice parameter of the ML $M - 1$. Therefore, the degree of elastic relaxation of the island, which increases from the bottom to the top of the island, induces a composition distribution that also increases from the bottom towards the top of the island.

7.4 Summary

In this chapter, we have presented a detailed TEM investigation of the structure and chemical morphology of free standing and capped In$_{0.6}$Ga$_{0.4}$As layers with nominal thicknesses of 1.5 and 2 nm as a function of the duration (0, 60 and 180 s) of a growth interruptions that was applied prior to the cap layer growth.

In uncapped samples, we found two kinds of islands: coherent islands with a diameter of approximately 13 nm, and large, plastically strain-relaxed islands. In the case of the coherent islands, the In concentration increases from the bottom to the top. In our opinion, the In distribution inside the islands is determined mainly by the segregation and strain-induced migration of In and Ga. Owing to the strain-induced migration, the In concentration of a growing ML M depends on the in-plane lattice parameter of the ML $M - 1$. We did not find indications of a WL in the regions beside the islands. This observation could be explained by a decomposition-induced depletion of the WL between the islands [12]. The strain field inside the islands allowed the WL thickness to be estimated as 1–2 ML during the very early stage of growth.

In capped samples, the density of coherent islands decreases (more quickly for the 2 nm sample) with increasing duration of the growth interruption. The density of dislocated islands increases. The chemical morphology of the capped samples significantly deviates from that of the uncapped samples. The structure of the In$_x$Ga$_{1-x}$As layer can be described by a quantum well about 4 nm thick with a (rather) homogeneous thickness, containing inclusions approximately 13 nm in size with an enhanced In concentration. The morphology transformation during the cap layer growth was explained by the interplay of three main effects, which are visualized in Fig. 7.13. Firstly, the unstable coherent islands are a source of In atoms, which are transported to the large dislocated islands via migration along the growth surface. The migrating In atoms are incorporated into the growing cap layer. Secondly, the strain-induced migration of Ga causes a significantly reduced growth rate of the GaAs cap layer on top of the elastically strain-relaxed islands. Therefore, the cap layer preferentially fills the regions between the islands. Thirdly, segregation occurs, leading to a floating layer of In on the growth surface. The measured concentration profiles yield a segregation probability $R = 0.810 \pm 0.006$ at a temperature of 500°C, which is in a good agreement with published data.

References

1. A. Rosenauer, W. Oberst, D. Litvinov, D. Gerthsen, A. Foerster, R. Schmidt: Phys. Rev. B **61**, 8276 (2000)
2. A. Rosenauer, T. Remmele, D. Gerthsen, K. Tillmann, A. Förster: Optik **105**, 99 (1997)
3. A. Ourmazd, P. Schwander, C. Kisielowski, M. Seibt, F.H. Baumann, Y.O. Kim: Inst. Phys. Conf. Ser. **134**: Section 1, 1 (1993)
4. C. Kisielowski, P. Schwander, F.H. Baumann, M. Seibt, Y.O. Kim, A. Ourmazd: Ultramicroscopy **58**, 131 (1995)
5. A. Rosenauer, D. Gerthsen: Adv. Imaging Electron Phys. **107**, 121 (1999)
6. Q. Xie, P. Chen, A. Madhukar: Appl. Phys. Lett. **65**, 2051 (1994)
7. O.M. Khreis, W.P. Gillin, K.P. Homewood: Phys. Rev. B **55**, 15813 (1997)
8. W.P. Gillin, D.J. Dunstan: Phys. Rev. B **50**, 7495 (1994)
9. G. Lu: Phil. Mag. Lett. **68**, 1 (1993)
10. H. Toyoshima, T. Niwa, J. Yamazaki, A. Okamoto: J. Appl. Phys. **75**, 3908 (1994)
11. H. Toyoshima, T. Niwa, J. Yamazaki, A. Okamoto: Appl. Phys. Lett. **63**, 821 (1993)
12. J. Tersoff: Phys. Rev. Lett. **81**, 3183 (1998)

8 InAs Quantum Dots

In the following, TEM and PL measurements which reveal mass transport and segregation in InAs SK layers grown on GaAs(001) by MBE at growth temperatures of 480°C and 530°C are presented. Plan-view TEM reveals regularly shaped islands with a density of 7.8×10^{10} cm^{-2} (480°C) and 1.5×10^{10} cm^{-2} (530°C). Uncapped islands were investigated by strain state analysis of electron wave functions in the image plane, reconstructed from HRTEM images. Indium concentration profiles of the islands were obtained by measurement of lattice parameter profiles of the islands and application of FE calculations. We found that the islands contained Ga atoms with an average concentration of 50% (480°C) and 67% (530°C). The capped InAs layers were investigated by PL and TEM. In agreement with TEM, PL indicates a smaller and deeper potential well for the islands grown at 480°C. Concentration profiles of the WLs were measured by TEM using the CELFA method, clearly revealing segregation profiles. The segregation efficiency of In atoms obtained is 0.77 ± 0.02 (480°C) and 0.82 ± 0.02 (530°C). As an explanation for the strong mass transport of Ga from the substrate into the islands, we suggest that the segregation of In atoms during the growth of the binary InAs can lead to the generation of vacancies in the metal sublattice. The vacancies penetrate from the WL and the islands into the GaAs buffer, leading to a unidirectional diffusion of Ga atoms from the buffer into the SK layer.

8.1 Experimental Setup

The epitaxial growth of the samples was carried out by G. Böhm at the Walter Schottky Institut, Technische Universität München, Garching. The samples were grown with a conventional MBE system using As$_4$. First, a GaAs buffer was deposited, followed by an AlAs/GaAs short-period superlattice (SSL) at a substrate temperature of 580°C, measured by an optical pyrometer. Then the growth was interrupted to reduce the substrate temperature to 530°C or 480°C, which was then kept constant during the growth of all further layers. After the growth of 30 nm GaAs, the deposition of InAs was started using a growth rate of 0.04 to 0.045 ML per second, depending on the specific sample, and an optimized arsenic pressure for each temperature. The symbols used to denote the samples studied are listed in Table 8.1, together with a summary of the corresponding growth conditions. During the growth of the InAs islands, the rotation of the substrate was stopped in order to vary the amount of deposited InAs on the wafer. The amount of InAs as a function of the position on the wafer was estimated assuming an In flux proportional to $\cos(\phi)/r^2$ [1]: the InAs amount varies almost linearly from -20% to $+20\%$ of the nominally deposited thickness along the diameter of a 2 inch wafer. The self-assembled InAs islands were capped with 30 nm GaAs, while the wafer was rotated. An AlAs/GaAs SSL was deposited on top to prevent carrier diffusion to the surface during PL measurements. The layer structure of the unburied samples was similar. For these samples, however, after the growth of the islands, the wafer was cooled down under arsenic pressure and then taken out of the MBE system.

The PL investigations were performed by M. Arzberger at the Walter Schottky Institut, Technische Universität München, Garching. The measurements were carried out at a temperature of 4.2 K. The 514 nm line of an Ar$^+$ laser was used for excitation. The spot size had a diameter of $\approx 200~\mu$m on the sample. The PL light was dispersed by a monochromator and detected by a liquid-nitrogen-cooled germanium detector using a standard lock-in technique. The PL spectra obtained were normalized with respect to the spectral intensity of the setup.

The TEM cross-section samples were prepared in the conventional way by Ar$^+$ ion milling at an energy of 3 keV in a liquid-nitrogen-cooled specimen holder. A Philips CM30 FEG/UT TEM equipped with an electron biprism, with a spherical-aberration constant $C_S = 0.65$ mm and a point-to-point resolution of 0.18 nm, was used. The images were recorded with an on-line CCD camera fixed at the exit of a Gatan imaging filter with 1024×1024 picture elements (pixels).

Table 8.1. Symbols used to denote the samples investigated, and the corresponding growth parameters [2], where "BEP" is the As$_4$ beam-equivalent pressure and t the nominal thickness of the InAs layer

	Sample			
	A	B	C	D
Cap layer	Yes	Yes	No	No
$T_G(°C)$	480	530	480	530
BEP GaAs (10^{-5} Torr)	1	1.3	1	1.3
BEP InAs (10^{-5} Torr)	0.15–0.17	1.2–1.3	0.15–0.17	1.2–1.3
t InAs (ML)	3.6 ± 0.3	2.7 ± 0.2	3.6 ± 0.3	2.7 ± 0.2

8.2 Experimental Results

8.2.1 Conventional Transmission Electron Microscopy

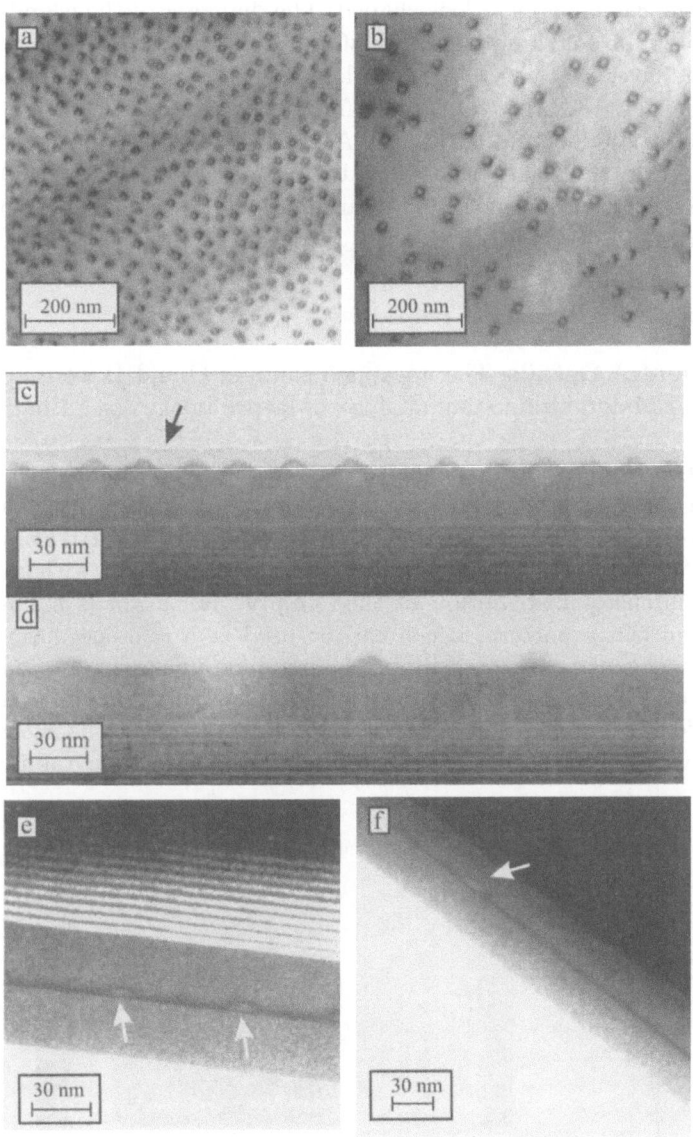

Fig. 8.1. TEM plan-view images of (**a**) sample A and (**b**) sample B. [100] cross-section TEM BF micrographs of (**c**) sample C and (**d**) sample D. (002) DF images of (**e**) sample A and (**f**) sample B

Figures 8.1a,b show plan-view micrographs of samples A and B. The islands appear as dark dots with a bright spot in the center. They are regularly shaped in both samples. Their density is 7.8×10^{10} cm^{-2} (sample A) and 1.5×10^{10} cm^{-2} (sample B). Figures 8.1c,d are zone-axis single-beam bright-field (BF) images taken from cross-sectional specimens of samples C and D. In both samples, the islands appear lens-shaped. The diameter of the islands at the bottom is ≈ 20 nm for sample C and ≈ 30 nm for sample D. The black arrow marks a gray layer that covers the island. This layer is a glue that was used to protect the islands during the TEM sample preparation. The bright lines at the bottom of the images indicate the AlAs layers of the AlAs/GaAs SSL. Figures 8.1e,f are DF images taken with the chemically sensitive (002) reflection. They give an overview of the WL (dark line) and the islands, which are marked by arrows.

8.2.2 Investigation of Uncapped Islands

8.2.2.1 Experimental Details The uncapped samples C and D were investigated by HRTEM and strain state analysis of lattice images (see Chap. 4).

The on-line CCD camera of the transmission electron microscope at the exit of the Gatan imaging filter causes geometric image distortions, which were corrected in the following way. Prior to the exposure of the HRTEM images, we took "empty" holograms using a Möllenstedt biprism without a specimen. The intensity distribution of the "empty" hologram is a homogeneously spaced fringe pattern, which can be used as a reference. Any deviations from homogeneity in the "empty" holographic image are caused by image distortions. Owing to image rotation in the transmission electron microscope, two "empty" holograms taken at different magnifications show fringe patterns that are rotated relative to each other. One pair of holographic

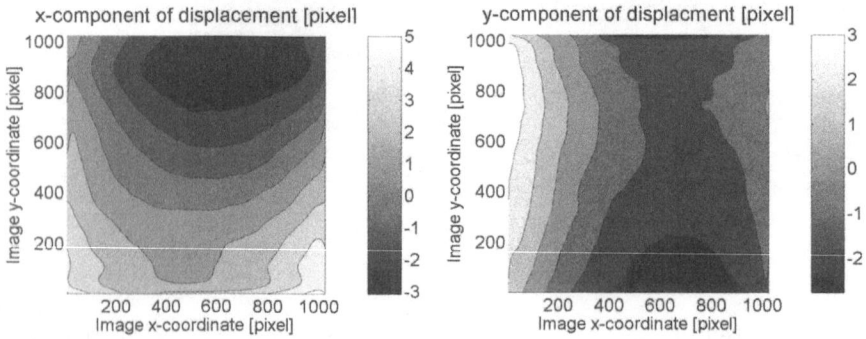

Fig. 8.2. Grayscale-coded maps of (*left*) the x component and (*right*) the y component of the displacement vector field caused by image distortions. The legends give the displacement in units of pixels

images was used to calculate a distortion map, which was used to correct the geometric distortion introduced by the Gatan imaging filter (and possibly by image distortions caused by the projector lenses). Figure 8.2 gives an example of a distortion map. The two graphs represent the 1024×1024 pixels of the CCD camera and show grayscale-coded maps of the x component (left graph) and y component (right graph) of the displacement vector field due to image distortions. The legends at the right of the graphs give the displacements in units of pixels.

To reduce the effects of objective lens aberrations such as the delocalization that leads to a broadening of interfaces in HRTEM images, strain state analysis was carried out by analyzing the amplitude images of the reconstructed wave function at the object exit surface, using the MAL method for focus-variation image reconstruction in HRTEM published in [3] (see Sect. 3.3.1). For that purpose, we took a series of 20 images with a defocus step size of 2.8 nm. We then corrected the distortion of each image of the defocus series and, finally, applied the MAL procedure using a program written by Coene et al. [3].

Additionally, we took holographic images of the investigated islands using a Möllenstedt biprism. The phase of the (000) beam in the sideband was used to estimate the specimen thickness (similarly to the method presented in [4]) close to the surface of the uppermost GaAs layer, and to estimate the wedge angle of the specimen caused by ion milling during the TEM sample preparation. We were not able to estimate the local thickness in the islands because they were embedded in an amorphous glue (see Fig. 8.3) that also caused a significant phase shift of the (000) beam of the sideband.

8.2.2.2 Details of Evaluation The reconstructed electron wave functions were used to evaluate local lattice parameters $a_{\parallel}$ parallel and $a_{\perp}$ perpendicular to the (001) plane, from which the concentration profile in the islands was deduced. Finally, our result was checked by FE calculations. Although the noise in the reconstructed images was significantly reduced by the MAL reconstruction compared with the noise level in the individual images of the defocus series, we applied an additional noise reduction using the Wiener filtering technique [5, 6] described in Sect. 4.1.1. In the fast-Fourier-transformed image, circular areas were chosen around the {022} reflections. Their radius was half the distance between the (000) and (022) reflections. The information outside the circles was deleted (except for the pixel in the center of the FFT). The result was inversely fast-Fourier-transformed. The amplitude image of the resulting complex image was used for further strain state analysis, which was carried out as described in Chap. 4. The reference region was chosen to be in the substrate, close to the lower left corner of the images displayed in Fig. 8.3.

8.2.2.3 Results Figure 8.3 shows three reconstructed amplitude images of samples C and D. The images clearly reveal the lens shape of the islands.

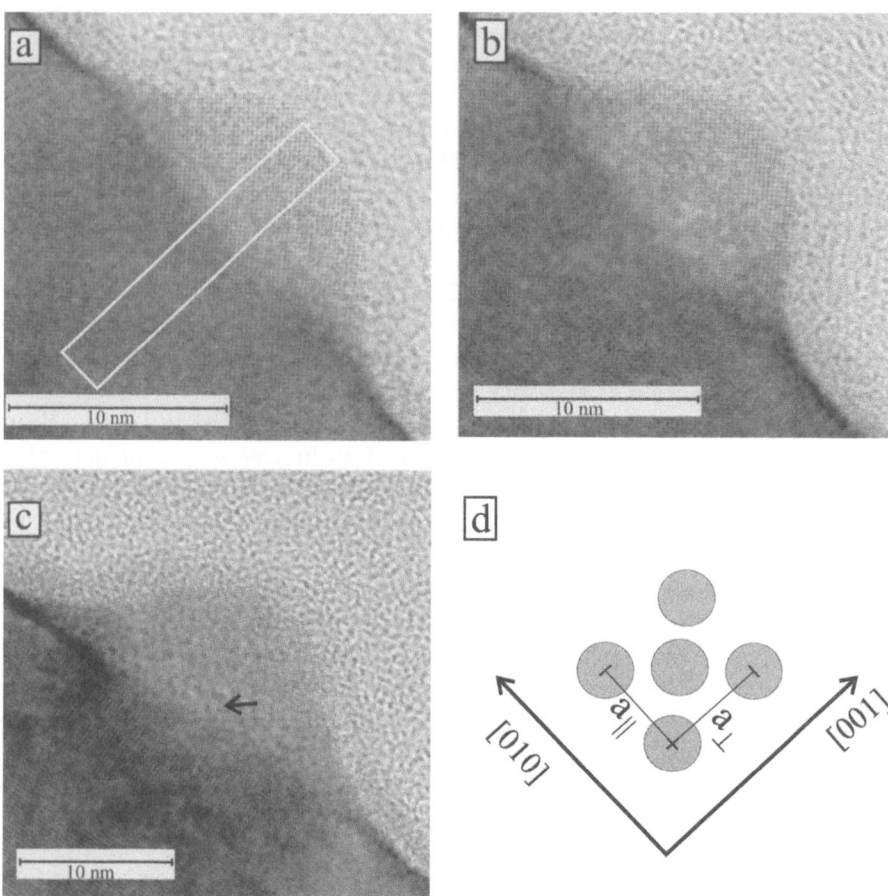

Fig. 8.3. Reconstructed amplitude images obtained by the MAL procedure [3] from defocus series of 20 [100] cross-sectional HRTEM images. (**a**) and (**b**) show different islands from sample C, and (**c**) corresponds to sample D. (**d**) Definition of the lattice parameters $a_{\parallel}$ and $a_{\perp}$. The *white frame* in (a) shows the area that was used to obtain averaged lattice parameters

Note that the scale of Fig. 8.3c is different from that of Figs. 8.3a,b. The islands in sample D (Fig. 8.3c) are significantly larger than those in sample C. Figure 8.4 shows the diameter of the islands plotted versus the distance in the [00$\bar{1}$] direction, given in units of ML, 0 corresponding to the top of the islands. Note that one ML here means the distance between two adjacent (002) lattice planes, and thus depends on the local (002) lattice plane spacing. Figure 8.4 shows that the height of the islands and their diameter islands at the bottom are larger for sample D.

We now outline the procedure that was applied to estimate the composition profile of an island. The local lattice parameters $a_{\parallel}$ and $a_{\perp}$ were averaged

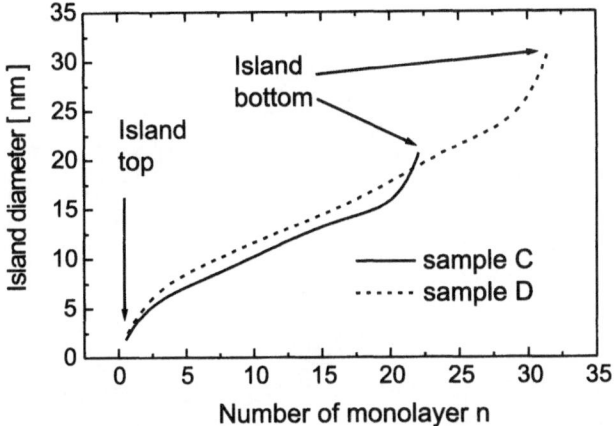

Fig. 8.4. Diameter of the islands plotted versus the distance in the [00$\bar{1}$] direction, given as a number of ML

along the [010] direction in a small region in the center of the island (white frame in Fig. 8.3a). Figures 8.5a,b show the resulting profiles of $a_\perp$ and $a_\parallel$ for sample C, obtained from the island shown in Fig. 8.3a. Appropriate analytical functions were applied to obtain fit curves for the experimental data. These were used to estimate the concentration profiles in the islands, as described in the following. Firstly, we assume that the composition of an island is laterally homogeneous. We show later that this assumption is valid to a good approximation. Secondly, we notice that the unit cells in the center of a rotationally symmetric island are biaxially strained. Therefore, the profiles of $a_\parallel$ and $a_\perp$ are correlated and reflect the biaxial strain state to a good approximation. We shall show the validity of this approximation later with the help of FE calculations. In this biaxial strain state, the lattice parameters $a_\parallel$ and $a_\perp$ of the strained crystal unit cell are correlated with the lattice parameter a of the unstrained cell according to

$$a_\parallel - a_\perp = \left\{ 1 + 2\frac{C_{12}(x)}{C_{11}(x)} \right\} (a_\parallel - a(x)) , \tag{8.1}$$

where the C_{ij} are elastic constants and x is the In concentration. We have used the bulk material lattice parameters and elastic constants listed in Table 4.1 and assumed that the elastic constants depend linearly on x. The dependence of the lattice parameter on the In concentration is given by Vegard's law as follows:

$$a(x) = xa(1) + (1 - x)a(0) . \tag{8.2}$$

Using the obtained fit curves for $a_\parallel$ and $a_\perp$, we solved (8.1) for x for each ML n of the profiles. The result is given in Fig. 8.5c for the islands 1 and 2, shown in Fig. 8.3a and b, respectively. Obviously, the two concentration

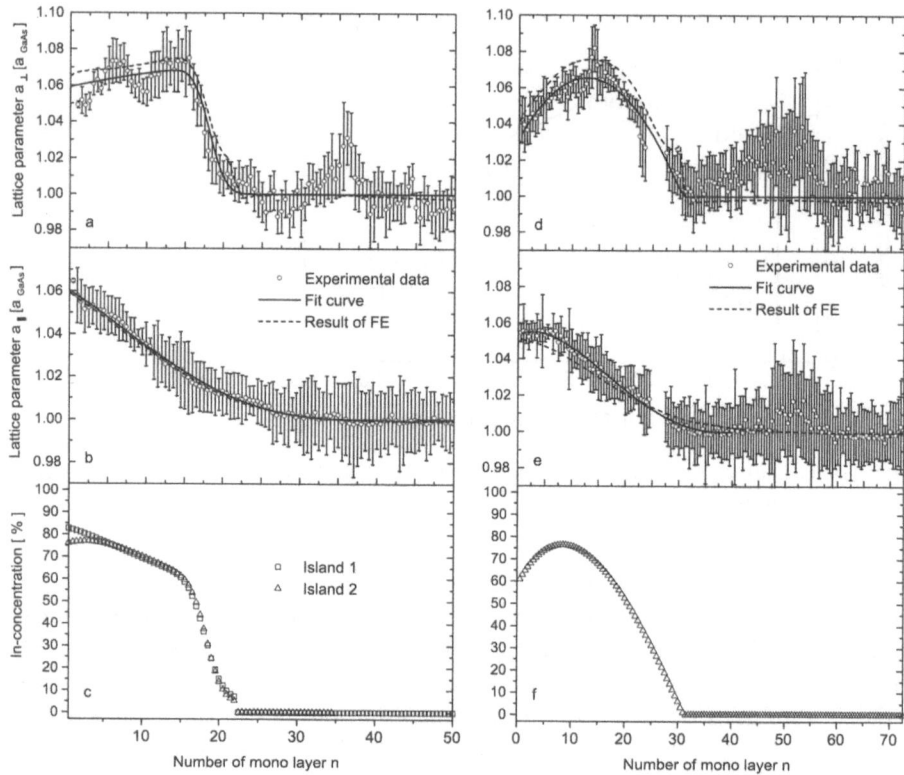

Fig. 8.5. (a),(b) Local lattice parameters $a_\perp$ and $a_\parallel$ obtained from sample C, plotted versus the distance in the [00$\bar{1}$] direction. The *left* margin corresponds to the top of the island. The values were averaged inside the white frame in Fig. 8.3a and the error bars give the corresponding standard deviation. The *solid curves* represent a fit of the experimental data and the *dashed* curves give the result of the FE calculation performed with the concentration profile shown in (c). (d), (e), (f) Analogous results obtained for sample D

profiles are in very good agreement, reflecting the homogeneity of the islands, also observed in Fig. 8.1a in capped samples. Figures 8.5d–f display the results for the island shown in Fig. 8.3c, corresponding to sample D. Figures 8.5c,f clearly reveal that the composition of the islands deviates significantly from the intended growth of binary InAs. Figure 8.5f reveals a decrease of the In concentration at the top of the island, which could be explained by desorption of In due to the larger growth temperature.

8.2.2.4 Comparison with FE Calculations We carried out FE simulations to check that the correlation of the profiles of $a_\parallel$ and $a_\perp$ reflected the biaxial strain state. We used (a) the assumption of a laterally homogeneous In concentration, (b) the concentration profiles shown in Figs. 8.5c,f, (c) a

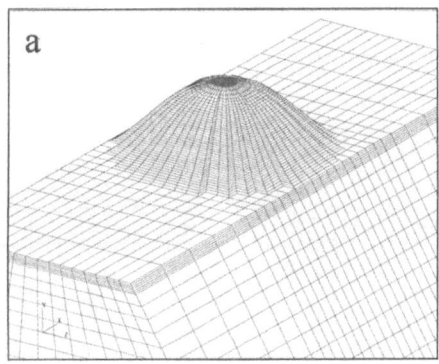 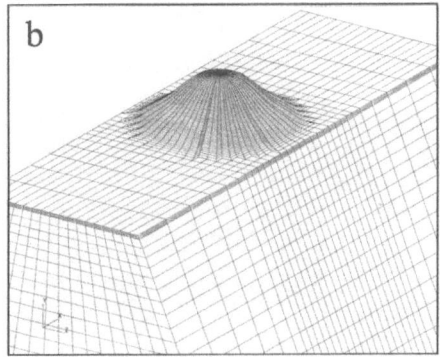

Fig. 8.6. (a), (b) FE models of the islands in samples C and D, respectively, used for the calculation of theoretical lattice parameter profiles

specimen geometry derived from the data obtained by electron holography (see Sect. 8.2.2.1) and the data shown in Fig. 8.4, (d) the assumption that the islands sit on top of the upper GaAs layer (and not on the WL), because the In distribution in the first few ML of the island differs from the composition of the surrounding WL (see Sect. 8.2.3). It was also assumed (e) that the islands were centered on the surface and (f) that there was a WL surrounding the islands that was consistent with the ascending part of the WL concentration profile measured in capped samples (see Sect. 8.2.3). However, the results of FE calculations carried out with a WL did not differ significantly from those obtained without a WL.

Figures 8.6a,b show the FE models used for the simulation of the islands in samples C and D. Note that the diameter of the island at its bottom is 22 nm in Fig. 8.6a and 32 nm in Fig. 8.6b. Each cell in Fig. 8.6 corresponds to a finite element. The height of the finite elements was half an ML in the islands and in the WLs. The FE simulations were carried out as described in Sect. 4.3.2, yielding the nodal displacement vectors **u** that give the deviations of the corners of the finite elements ("nodes") from their original positions. These original positions are equivalent to the reference lattice obtained from the GaAs in the experimental image. After the calculation of the nodal displacements, they were averaged along the "electron beam direction", and we obtained a map of the projected displacement vectors. The result was used to evaluate the profiles of $a_{\parallel}$ and $a_{\perp}$ in the central parts of the islands, in analogy with the evaluation of the experimental images. The results are shown as dashed lines in Figs. 8.5a,b,d,e. Obviously, the agreement is excellent for Figs. 8.5a,b and still fair for Figs. 8.5d,e. The small deviation between the FE simulation and the experimental data observed in Figs. 8.5d,e is most likely caused by a truncation of a small part of the island during the ion milling in the TEM sample preparation.

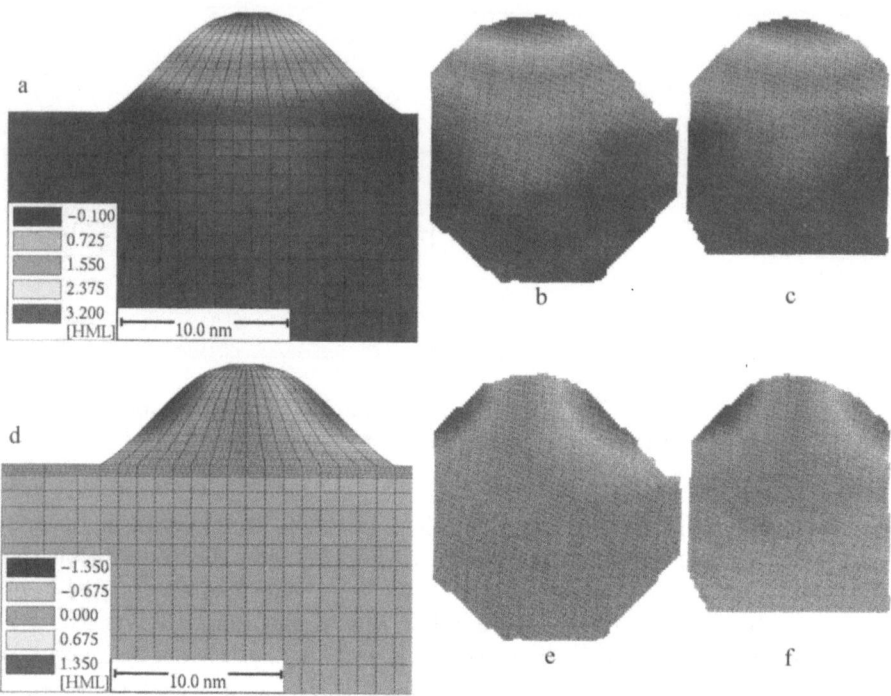

Fig. 8.7. Color-coded maps of displacement vector components. (**a**),(**d**) Results of the FE calculation. The *upper row* of images shows the displacement vector components in the [001] growth direction and the *lower row* shows the components in the [010] direction (parallel to the surface). The legends give the displacement components in units of half an ML (HML). The color coding is kept constant for each of the two rows. (**b**), (**e**) Results for the island in sample C depicted in Fig. 8.3a. (**c**),(**f**) Maps of the island shown in Fig. 8.3b [Please see Plate 23 for color reproduction of this plate]

Figure 8.7 compares the simulated displacement vector fields with the experimental results obtained for two islands in sample C. The upper row of images displays color-coded maps of the displacement vectors projected onto the [001] direction, and the lower row corresponds to the [010] components. The scaling is kept constant for each of the two rows. Figures 8.7a–c show that the displacements vanish in the GaAs layer, whereas they increase from the bottom to the top of an island owing to its larger lattice parameter. Figures 8.7d–f show that the displacement vectors close to the side faces of an island point outwards and thus give evidence for the elastic relaxation of the island. Similar results are shown for sample D in Figs. 8.8a,c (FE simulation) and b,d (evaluation of the island shown in Fig. 8.3c). The white rectangle in Fig. 8.8b is due to am image pattern that was not sufficiently pronounced, marked by a

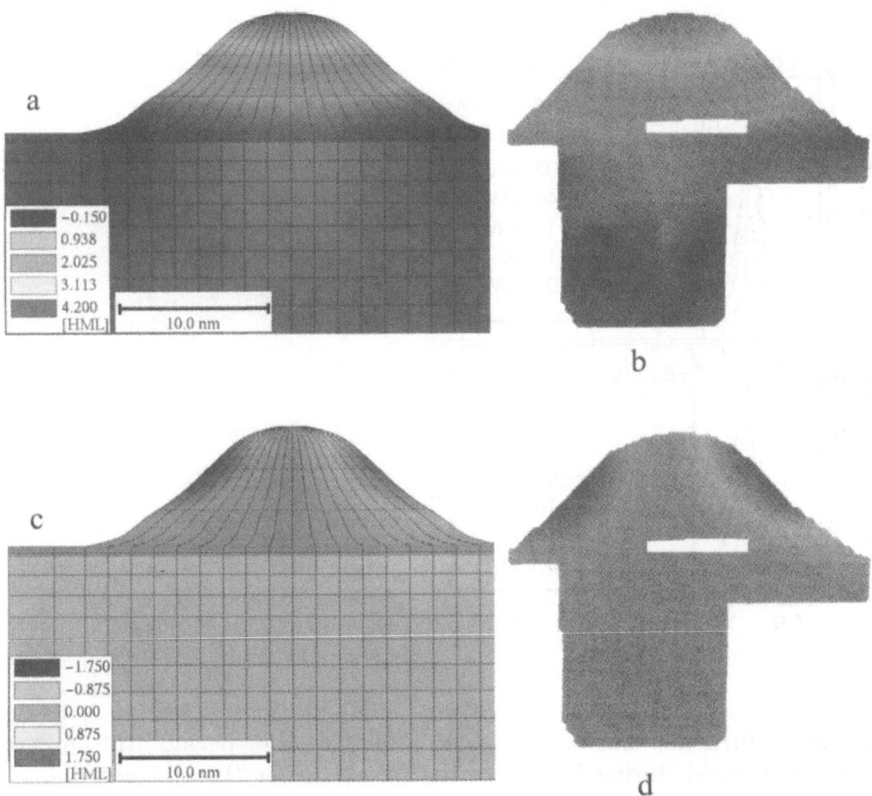

Fig. 8.8. Color-coded maps of projected displacement vectors, analogous to those
in Fig. 8.7. (b),(d) Results for the island depicted in Fig. 8.3 (sample D) [Please
see Plate 24 for color reproduction of this plate]

black arrow in Fig. 8.3c. In this region, the atomic positions cannot be clearly
resolved and thus were not evaluated. This is also the reason for some missing
data points in Figs. 8.5d,e close to $n = 25$. Nevertheless, Fig. 8.8 reveals that
the In concentration profile in Fig. 8.5f leads to consistent results for the
simulated and evaluated displacement vector fields over the whole (projected)
island. The good agreement between the displacement fields obtained by FE
simulation and by evaluation of reconstructed HRTEM images also indicates
that the assumption of a laterally homogenous composition in the islands is
valid to a good approximation.

8.2.3 Investigation of Capped Samples

8.2.3.1 PL Figure 8.9 displays PL spectra of the two capped QD samples,
grown at substrate temperatures of 530°C (sample B) and 480°C (sample
A) measured at 4.2 K with various excitation powers. As the amount of

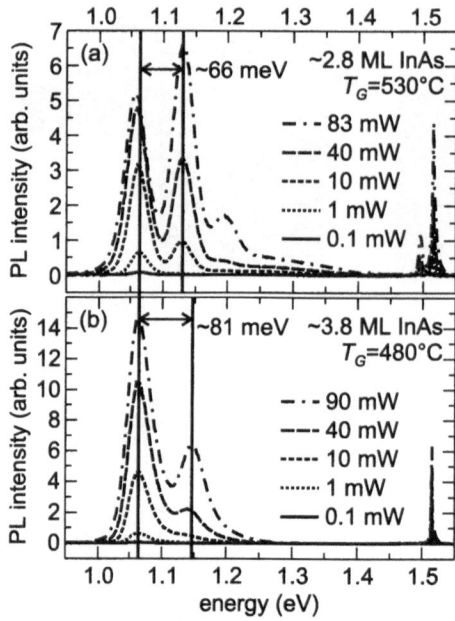

Fig. 8.9. PL spectra of (a) sample B and (b) sample A, measured with various excitation powers at 4.2 K

deposited InAs depends on the position on the wafer because the substrate was not rotated during InAs growth, the spots for the PL measurements could be chosen in order to obtain the same ground-state transition energy for both samples: the estimated amounts of deposited InAs at the positions investigated are ~2.8 ML and ~3.8 ML at 530°C (Fig. 8.9a) and 480°C (Fig. 8.9b), respectively. Since the density of the islands is increased by about five times when the substrate temperature is reduced from 530°C to 480°C, much more InAs must be deposited to achieve a similar amount of InAs per QD and therefore a similar ground-state transition energy at the lower growth temperature. The FWHM of the ground-state transition for both samples is 35–40 meV and shows the good homogeneity of the islands at both growth temperatures. When the excitation power is increased, the ground state intensity rises and luminescence from excited states is observed for both samples. The ground-state transition intensity of the 530°C QDs saturates at the excitation densities used (Fig. 8.9a), and the luminescence intensity from excited states exceeds the ground-state transition. As the electron–hole pairs were photo generated in the GaAs by an argon laser beam (514 nm), the number of carriers per QD is inversely proportional to the island density, assuming that all generated electron–hole pairs relax into the QDs. The QD carrier density, therefore, is about five times higher for the 530°C QDs than for the 480°C QDs at the same excitation power. Hence the spectra of the 480°C QDs are dominated by the ground-state transition for all excitation densities used (see Fig. 8.9b). The maximum ground-state intensity is about three times higher than the saturated ground-state intensity of the 530°C QDs due to the higher island density. The peak splitting, i.e. the difference between the energies of the first excited-state transition and the ground-state transition, is ~66 meV for the QDs grown at 530°C. Although the ground-state energy is the same for the 480°C QDs, we observe a significantly larger peak splitting of ~81 meV for these QDs.

8.2.3.2 Investigation of the WL by TEM

Brief Outline of the Evaluation Technique. For the evaluation of the composition profiles of the WL in capped samples (A and B), we used the CELFA technique [6, 7, 8] described in Chap. 5. An interference of the (002) beam with the (000) beam was chosen for the image formation. Figure 8.10a shows an example of a fringe pattern obtained from sample A.

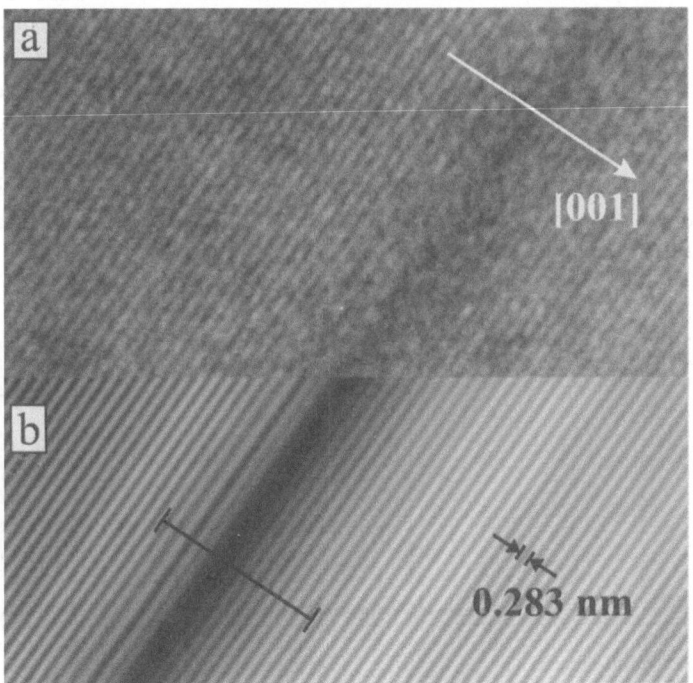

Fig. 8.10. (a) Lattice fringe image of the WL of sample A. The fringes correspond to (002) lattice planes. (b) Noise-reduced lattice image obtained by averaging along fringes. For the averaging, we used a Gaussian-shaped weighting function with an FWHM value of typically 200 pixels

To avoid noise reduction that could introduce a small broadening of the measured profiles, we performed an averaging of each lattice fringe image parallel to the fringes prior to the evaluation. This technique also improves the signal-to-noise ratio but does not reduce the information content of the image in the growth direction (perpendicular to the lattice fringes). Of course, the image resolution in the direction parallel to the fringes is decreased. This effect is not crucial if the image is homogeneous along the fringes. Here, this condition is well fulfilled because we are valuating images of the WL that are laterally homogeneous to a good approximation. The result of the averaging is shown in Fig. 8.10b.

For the CELFA evaluation, a specimen thickness of 15 nm was assumed. This assumption does not significantly reduce the accuracy, because the maximum error in the concentration evaluation is $\Delta x = 2\%$, calculated for the range $0 < t \leq 40$ nm and $0 \leq x \leq 50\%$ (see Sect. 5.4 and Fig. 5.23).

Note that the (002) lattice planes are distorted by strain because the lattice parameter of InAs is larger than that of GaAs. Important effects include the tetragonal distortion of the (biaxially) strained WL and its partial strain relief due to the small TEM specimen thickness. This effect can reduce the biaxial strain state to a uniaxial one for an infinitely small specimen thickness and leads to lattice plane bending close to the surfaces of the specimen. To estimate the strength of strain effects, FE and MS simulations were carried out as described in Sect. 5.4. There, we found that the effect of lattice parameter variation and lattice bending on the amplitude of the (002) beam was rather small, leading to a maximum error $\Delta x < 5\%$ for a 5 nm thick $In_{0.5}Ga_{0.5}As$ layer covered with 10 nm GaAs. The error is even less for a smaller total content of indium or a smaller In concentration.

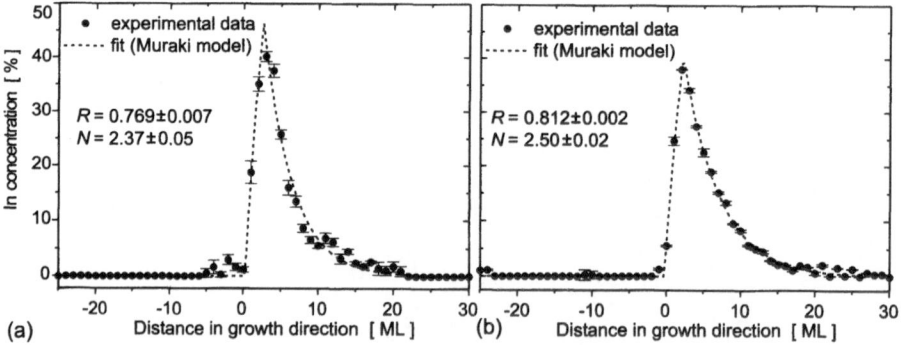

Fig. 8.11. Profiles of the In concentration in the WL of (**a**) sample A and (**b**) sample B, obtained by CELFA. The *right-hand side* corresponds to the GaAs cap layer. The *error bars* give the standard deviation, obtained by averaging over the image area. The *dashed lines* show a fit of the experimental data obtained with the model of Muraki et al. [9]

Results. Figure 8.11 displays examples of the resulting In concentration profiles for the capped samples A and B. The profiles clearly show the characteristics of segregation: a steep increase from the GaAs buffer layer towards the WL and an exponential decrease towards the cap layer. The concentration profiles were fitted by use of the Muraki model of segregation [9], given by (6.6). The segregation probability R and the number N of deposited ML were used as fit parameters, and the nominal In concentration x_0 was set to 1. The values of the fit parameters are given in Fig. 8.11. Obviously, both profiles can be fitted well by the Muraki model. The profile of sample A, grown at 480°C, is somewhat sharper ($R = 0.77$) than that of sample B ($R = 0.81$), grown at the higher temperature of 530°C. Table 8.2 gives an overview of the data obtained from a variety of different evaluated images. The second and third columns show the values obtained for the fit parameters N and R. The fourth column gives the content of In in the exponentially decreasing part of the segregation profile, i.e. the In content in the surface layer given by (6.7).

Table 8.2. Data obtained for the WLs in sample A ($T_G = 480°C$) and B ($T_G = 530°C$)

Growth temperature	N (ML)	Segregation probability R	x_{surf}^{In} (ML)
480°	2.3±0.2	0.77±0.02	1.5±0.2
530°	2.5±0.2	0.82±0.02	1.8±0.2

8.2.4 Combination of the Results

Here, we combine the results described above to gain a complete picture of the structure of the InAs SK layers. The In content of an island, in units of ML InAs, is given by

$$C_{In}^{Isl} = \frac{\sum_h \pi[d(h)/2]^2 x_{In}(h)\,\Delta h}{A} \,, \tag{8.3}$$

where h is an index that numbers the ML from the bottom to the top of the island, $d(h)$ is the diameter of the island as a function of the height (see Fig. 8.4), $x_{In}(h)$ is the In concentration profile (see Figs. 8.5c,f), $\Delta h = 1$ ML is the thickness of an ML, and A is a "reference" area. An analogous equation is used for the Ga content. For $A = \pi(d(0)/2)^2$, we obtain an In content referred to the basal surface of an island, listed in the third column of Table 8.3. The fourth column gives the value "per island area", i.e. we have used $A = 1/\delta$, where δ is the island density. The fifth column of Table 8.3 contains the In content of the WL, calculated according to

$$C_{\text{In}}^{\text{WL}} = N\frac{1/\delta - \pi\,[d(h)/2]^2}{1/\delta} \ , \tag{8.4}$$

where N is the fit parameter of the Muraki model of segregation. The values of N are listed in Table 8.2. From these data, we are able to calculate the deposited amount of InAs by summing columns 4 and 5 in Table 8.3. The result is given as the "total" In content in column 6. Column 7 lists the nominally deposited amounts of InAs. Obviously, columns 6 and 7 are in good agreement.

Table 8.3. Contents of In and Ga in the Stranski-Krastanov layers given in units of ML InAs and GaAs. The values for the islands were obtained from the uncapped samples C, and D, whereas the data for the WL were obtained from the capped samples A and B

T_G	Material	Per basal surface	Per island area	WL	Total	Nominal
480°	In	9	1.6	1.9±0.2	3.5±0.2	3.6±0.3
	Ga	9	1.6			
530°	In	7.4	0.8	2.2±0.2	3.0±0.2	2.7±0.2
	Ga	11.3	1.2			

8.3 Discussion

8.3.1 Correlation Between TEM and PL Data

The results obtained by TEM for the uncapped samples C and D show that the islands grown at a substrate temperature of 480°C are smaller and contain more indium than those grown at 530°C (Figs. 8.3 and 8.5). Hence, the confinement potential for carriers created by the islands after the overgrowth is deeper but narrower for the 480°C QDs. The two samples investigated by PL (Fig. 8.9) were chosen in order to obtain the same ground-state transition energy. So, we have to consider two confinement potentials with the same ground-state energy, but one of them is deeper and narrower. As sketched schematically in Fig. 8.12, this leads in principle to an increased splitting between the ground-state energy E_0 and first excited-state energy E_1 for the deeper confinement potential. The increased peak splitting between the ground-state transition and the first excited-state transition observed in PL for the 480°C QDs, therefore, can be explained by different confinement potentials, which depend on the growth temperature. This result is consistent with the TEM data, which showed a smaller island height but a larger In concentration and a larger In content (see Table 8.3) in sample C than in sample D.

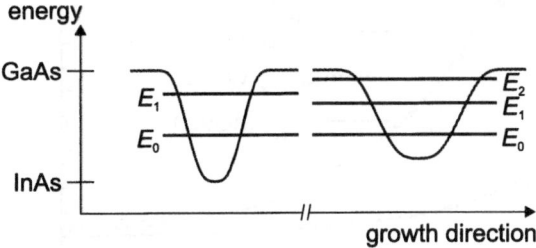

Fig. 8.12. Schematic drawing of two confinement potentials with the same ground-state energy E_0. The *left* potential well is deeper but narrower, leading to a higher energy E_1 of the first excited state

8.3.2 Segregation Efficiency

The concentration profiles of the WLs in samples A and B (see Fig. 8.11) clearly indicate the effect of segregation. Both of the profiles in Fig. 8.11 can be fitted well by the model of segregation (6.6) suggested in [9]. Note that we were not able to fit our data with (6.5). The evaluated segregation efficiencies (see Table 8.2) increase with increasing growth temperature. This behavior and the values obtained for R are consistent with published data for in ternary InGaAs. Muraki et al. [9] obtained $R \approx 0.85$ ($T_G = 500°C$, $x_0 = 0.126$), Toyoshima et al. [10] observed $R = 0.81 - 0.85$ ($T_G = 550°C$, $x_0 = 0.15 - 0.31$, As_4 pressure 1.4×10^{-5} Torr, InAs growth rate 0.09 ML/s), and also [11] $R = 0.71 - 0.76$ ($T_G = 475°C$, $x_0 = 0.34$, As_4 pressure $5.6 - 1.3 \times 10^{-5}$ Torr, growth rate 0.09 ML/s) and $R = 0.84$ ($T_G = 520°C$, $x_0 = 0.34$, As_4 pressure 1.3×10^{-5} Torr, InAs growth rate 0.09 ML/s). The values given by Kaspi and Evans [12] show a linear dependence on the growth temperature and range from $R \approx 0.68$ at $T_G = 425°C$ to $R \approx 0.83$ at $T_G = 520°C$ ($x_0 = 0.22$, As_2 dimer equivalent pressure 5×10^{-6} Torr, growth rate 0.97 ML/s). In [13], $R = (0.81 \pm 0.006)$ is given ($T_G = 500°C$, $x_0 = 0.6$).

8.3.3 Segregation Models

The publications listed in the previous section and the investigation of WLs presented here indicate that segregation is well described by the model of Muraki et al. [9] even if the "surface" content x^{In}_{surf} of InAs is larger than 1 (see Table 8.2). It is important to note that this cannot be achieved according to the model that was suggested by Moison et al. [14], which uses (6.4). As a consequence, both models can be used to describe experimental

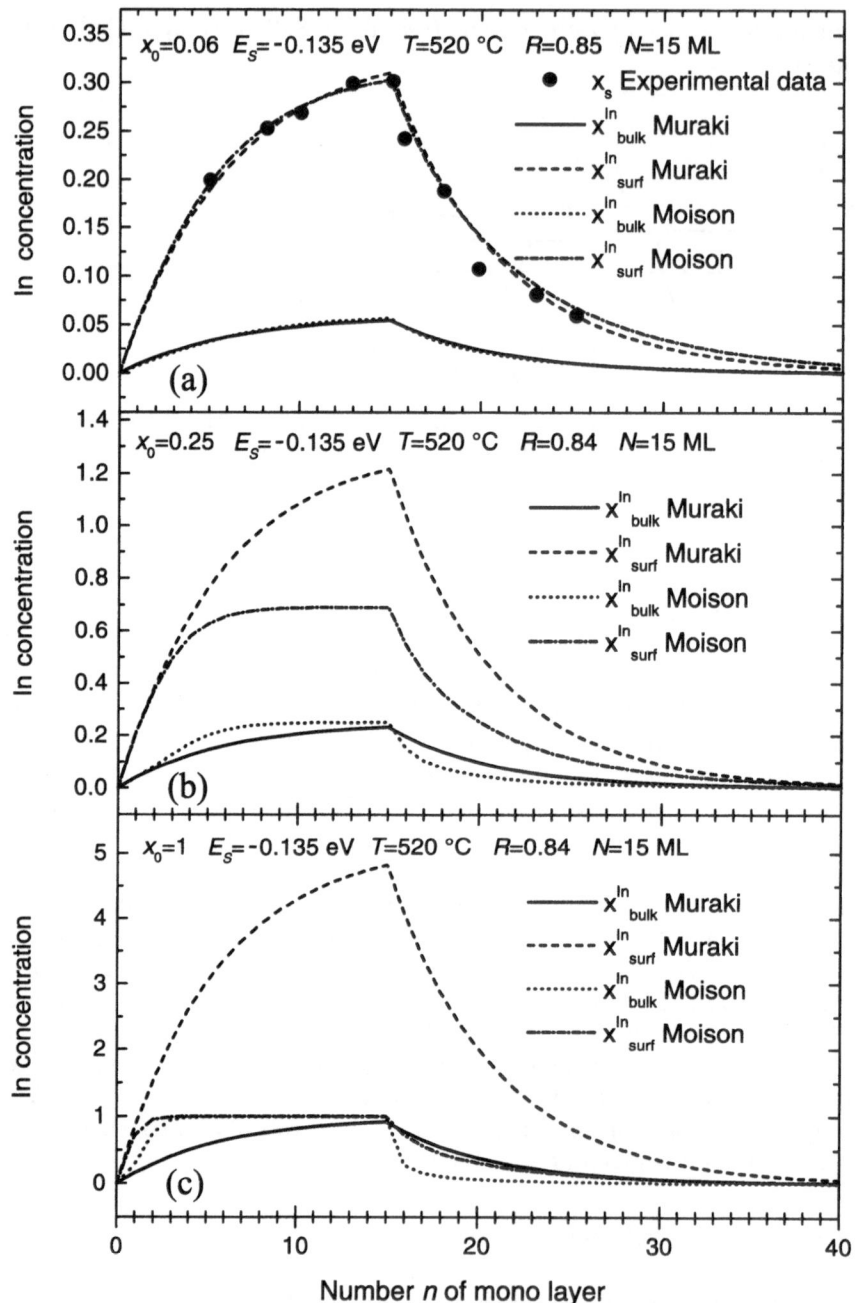

Fig. 8.13. Calculated profiles of the In concentration x^{In}_{bulk} in the bulk layer and x^{In}_{surf} in the surface layer according to the segregation models suggested by Muraki et al. [9] and Moison et al. [14], with the parameters given. (a) contains experimental data x^{In}_{surf} from [15]

data for small nominal In concentrations x_0 but they deviate for larger x_0 if $x_{\text{surf}}^{\text{In}} > 1$. Figure 8.13 visualizes this consideration more clearly. In Fig. 8.13a, we show the result of fitting the experimental data for $x_{\text{surf}}^{\text{In}}$ published by Gerard et al. [15] with the model suggested by Muraki et al. [9] (6.7) and with that given by Moison et al. [14] (6.5). Both models yield a satisfactory fit of the experimental data. Keeping the fit parameters obtained (x_0, the segregation energy E_S and R), we find an increasing deviation between the In concentration profiles $x_{\text{bulk}}^{\text{In}}$ of the two models with increasing x_0 (0.25 in Fig. 8.13b and 1 in Fig. 8.13c). This deviation is caused by an In content $x_{\text{surf}}^{\text{In}}$ that becomes larger than 1 in the case of the Muraki model, which cannot occur if we use the Moison model, because (6.5) combined with the condition $x_{\text{Ga}} \geq 0$ yields $x_{\text{surf}}^{\text{In}} \leq 1$.

We now show that the Muraki model, under certain conditions, implicitly violates $x_{\text{Ga}} \geq 0$. For that purpose, we "reconstruct" the Ga concentrations of the individual ML as shown in Fig. 8.14. The right column in Fig. 8.14 gives the In concentrations obtained from (6.6) and (6.7) during the growth of the first two ML. The left column in Fig. 8.14 shows the corresponding Ga concentrations. All concentration values of an element M given here have to be understood as the number of atoms of type M (per ML) divided by the number of available lattice sites in the metal sublattice (per ML). First, one ML of $\text{In}_{x_0}\text{Ga}_{1-x_0}\text{As}$ is deposited. In the following segregation step, a fraction Rx_0 of In atoms segregates to the next layer, where the Ga concentration is 0. Then, the second ML is deposited. At the surface, the Ga concentration is $1 - x_0$ and the In concentration is $x_0 R + x_0$. The concentration of occupied lattice sites in the underlying layer is $(1 - x_0) + x_0(1 - R) = 1 - x_0 R$. To obtain a completely filled metal sublattice without changing the distribution of In atoms, we eliminate the vacancies with Rx_0 Ga atoms from the surface layer. In the following segregation step, a third layer is generated without Ga atoms and with an In concentration of $(x_0 R + x_0)R$. We now focus on the Ga concentration x_{Ga} in the subsurface ML that develops after the growth of each ML n (bold boxes in Fig. 8.14). We find

$$x_{\text{Ga}}(n) = \begin{cases} 1 - x_{\text{surf}}^{\text{In}}(n) - x_0 : 1 \leq n \leq N \\ 1 - x_{\text{surf}}^{\text{In}}(n) \quad\quad : n > N \end{cases} \tag{8.5}$$

Obviously, the condition $x_{\text{Ga}} > 0$ is violated if $x_{\text{surf}}^{\text{In}}(n - 1) > 1 - x_0$. Figure 8.15 demonstrates this behavior for parameters N, R and x_0 corresponding to Figs. 8.13a,b,c, respectively. It is obvious that $x_{\text{Ga}} > 0$ is obeyed for $x_0 = 0.06$, whereas it is violated for $x_0 = 0.25$. In summary, these considerations show that each In atom that moves from the upper crystal layer to the floating layer leaves behind a vacancy. This vacancy can be filled by a Ga atom during the growth of the subsequent layer if there are sufficient Ga atoms available. On the other hand, the vacancy remains if there is no Ga atom available. In the model described above, this is the case if $x_{\text{Ga}} < 0$. In this case, the number of available Ga atoms is not sufficient to fill all empty sites in the

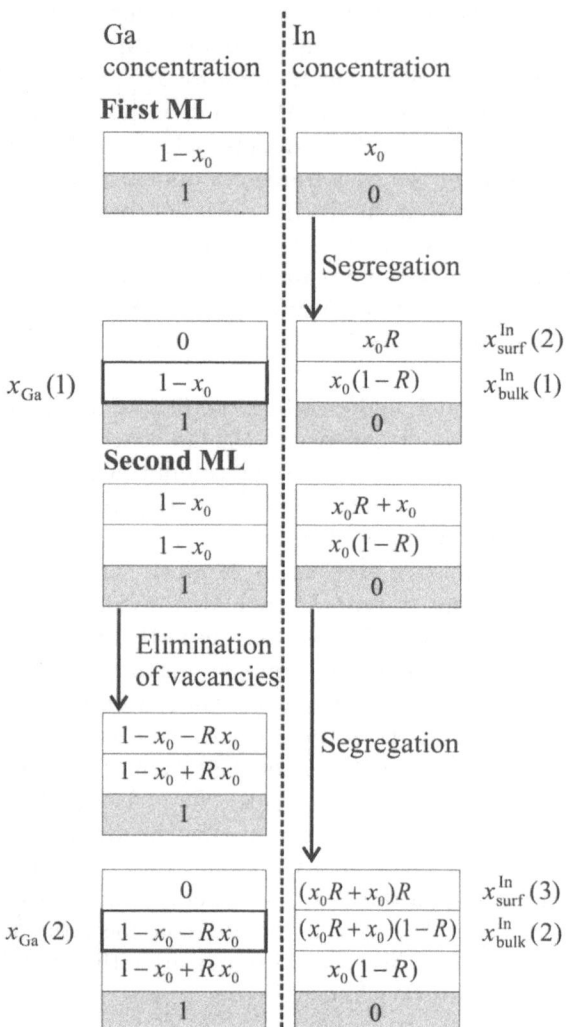

Fig. 8.14. Sketch showing the concentrations of In and Ga atoms relative to the lattice sites that are available in the metal sublattice, during the growth of the first two ML of $In_{x_0}Ga_{1-x_0}As$. Each box corresponds to 1 ML. The *right column* contains the In concentrations calculated according to the segregation model suggested by Muraki et al. [9]. The *left column* gives the Ga concentrations obtained as described in the text

metal sublattice, and vacancies are left behind. Note that filling the remaining empty sites in the metal sublattice with In atoms from the floating layer or from the impinging molecular beam would lead to a deviation of x_{surf}^{In} from the profile given by the Muraki model in (6.7).

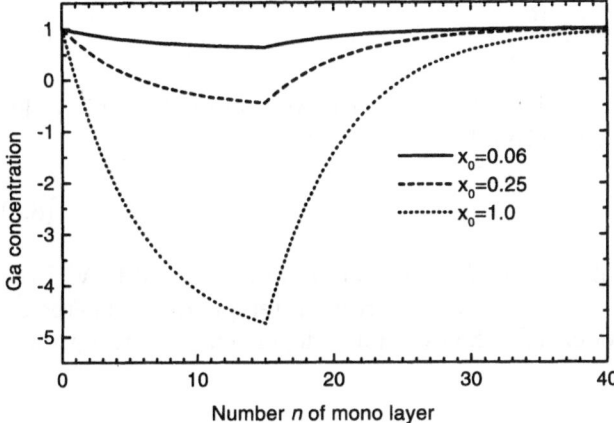

Fig. 8.15. Ga concentration profiles corresponding to Figs. 8.13a,b,c, respectively. The profiles were calculated from (8.5) and reveal the implicit violation of $x_{Ga} > 0$ in the Muraki segregation model [9]

8.3.4 Modeling

Since the segregation model suggested by Muraki et al. [9] describes the experimental data well, we have to conclude that (under the conditions described above) the segregation of In leads to the formation of vacancies in the metal sublattice. However, it seems likely that the vacancies are not stable but anneal out during layer growth. One possible process is the diffusion of vacancies into the GaAs buffer layer, so that a vacancy is filled with either an In atom or a Ga atom from the underlaying layer.

At this point, the difference between the segregation model of Muraki et al. [9] and that of Moison et al. [14] (6.5) becomes clearer. The Moison model is based on an exchange reaction of In and Ga. This model is formulated as a mass-action law that obeys (6.4). In contrast, the Ga atoms are not considered in the Muraki model. The Muraki model provides a better description of the experimental data because the Ga atoms seem to be delivered by different processes, such as diffusion of vacancies, as discussed above. It is interesting to see that the Moison model becomes identical to the Muraki model if the Moison model is modified in such a way that it does not take the Ga atoms into account. We consider the reaction of an In atom moving between the bulk-like bonding of the last grown ML of the specimen and the floating layer of physisorbed In atoms on top of the specimen as described in Sect. 6.2. The mass-action law of (6.3) becomes

$$-\frac{E_S}{kT} = \ln x^{In}_{surf} - \ln x^{In}_{bulk} , \qquad (8.6)$$

which we compare with the Muraki model and obtain

$$\exp\left(-\frac{E_{\mathrm{S}}}{kT}\right) = \frac{x_{\mathrm{surf}}^{\mathrm{In}}}{x_{\mathrm{bulk}}^{\mathrm{In}}} = \frac{R}{1-R} \, , \tag{8.7}$$

where R is the segregation efficiency used in the Muraki model. This yields a relation between R and E_{S} given by

$$E_{\mathrm{S}} = kT \ln \frac{1-R}{R} \, . \tag{8.8}$$

For a temperature $T = 520°\mathrm{C}$ and $R = 0.8$, we obtain $E_{\mathrm{S}} = -0.1$ eV. Note that (8.8) predicts a decrease of R with increasing temperature. Therefore, the modified Moison model also shows a false dependence on the growth temperature.

8.3.5 Segregation and Mass Transport of Ga

In the preceding sections we suggested that the segregation of In leads to the generation of vacancies and thus to a mass transport of Ga atoms for sufficiently large x_0. Segregation of In atoms can also be expected during the growth of the islands and thus should lead to a transport of Ga atoms from the GaAs buffer into the islands. Therefore, the high content of Ga that was observed inside the islands as described in Sect. 8.2.2 can be explained by the mass transport of Ga due to segregation. Assuming mass transport by diffusion of segregation-induced vacancies into the GaAs buffer layer, the number of vacancies that pass through the basal plane of an island depends on the island height. Therefore, the amount of Ga close to the basal plane increases with the height of the island. This effect could explain the steeper increase of the In concentration in the islands in sample C (Fig. 8.5c) compared with sample D (Fig. 8.5f).

Khreis et al. found that the diffusion coefficient of vacancies in the metal sublattice is five orders of magnitude larger than that of In in GaAs [16]. Extrapolating the vacancy diffusion coefficients obtained between 750°C and 890°C given in [17], we obtain a diffusion length of only ≈ 1 ML during the growth of 1 ML InAs. On the other hand, the extrapolation of high-temperature coefficients can be rather inaccurate. In addition, the measured values of diffusion cofficients of In in GaAs deviate from one another by orders of magnitude (compare the values given in [18, 16, 19], for example). Leon et al. [19] performed annealing experiments on $\mathrm{In}_{0.5}\mathrm{Ga}_{0.5}\mathrm{As}/\mathrm{GaAs}$ quantum wells. They found increased diffusivity values at short annealing times due to impurity interaction. This could be explained by a segregation-induced increase of the vacancy concentration in the vicinity of the $\mathrm{In}_{0.5}\mathrm{Ga}_{0.5}\mathrm{As}$ layers.

The effect of mass transport of Ga atoms from the buffer layer into the islands has been observed by TEM [20], scanning tunneling microscopy [21] and surface-sensitive X-ray diffraction [22]. In [21], Joyce et al. measured the volumes of InAs QDs on GaAs(001) and found that the volume of the QDs can exceed the total amount of deposited InAs. These authors concluded that the dots must contain both Ga and In, with a Ga fraction of about 30% (T_G was 500°C, 3 ML InAs were deposited).

8.4 Conclusion

In conclusion, we have investigated InAs QDs grown at two different temperatures $T_G = 480$°C and 530°C. By TEM, we found a gallium fraction of 50% (480°C) and 67% (530°C) in islands in uncapped samples. The results were in good agreement with PL data obtained from capped samples. In capped samples, our TEM investigation of the WL revealed segregation profiles with a segregation efficiency of indium $R = 0.77 \pm 0.02$ (480°C) and 0.82 ± 0.02 (530°C), which is in a good agreement with published data obtained from nominally ternary InGaAs. By measurement of island densities and geometry, and of composition profiles in islands and WLs, we calculated the amount of deposited InAs was in agreement with the nominal amount. Mass transport of Ga atoms from the GaAs buffer into the islands was discussed in connection with a segregation-induced generation of vacancies in the metal sublattice.

References

1. I. Brodie, J.J. Muray: *The Physics of Microfabrication*, 1st edn. (Plenum, New York, 1987)
2. L. Chu, M. Arzberger, G. Böhm, G. Abstreiter: J. Appl. Phys. **85**, 2355 (1999)
3. W.M.J. Coene, A. Thust, M. Op de Beeck, D. Van Dyck: Ultramicroscopy **64**, 109 (1996)
4. A. Rosenauer, D. Van Dyck, M. Arzberger, G. Abstreiter: Ultramicroscopy **88**, 51 (2001)
5. A. Rosenauer, S. Kaiser, T. Reisinger, J. Zweck, W. Gebhardt, D. Gerthsen: Optik **102**, 63 (1996)
6. A. Rosenauer, D. Gerthsen: Adv. Imaging Electron Phys. **107**, 121 (1999)
7. A. Rosenauer, U. Fischer, D. Gerthsen, A. Förster: Ultramicroscopy **72**, 121 (1998)
8. A. Rosenauer, D. Gerthsen: Ultramicroscopy **76**, 49 (1999)
9. K. Muraki, S. Fukatsu, Y. Shirakia, R. Ito: Appl. Phys. Lett. **61**, 557 (1992)
10. H. Toyoshima, T. Niwa, J. Yamazaki, A. Okamoto: Appl. Phys. Lett. **63**, 821 (1993)
11. H. Toyoshima, T. Niwa, J. Yamazaki, A. Okamoto: J. Appl. Phys. **75**, 3908 (1994)
12. R. Kaspi, K.R. Evans: Appl. Phys. Lett. **67**, 819 (1995)
13. A. Rosenauer, W. Oberst, D. Litvinov, D. Gerthsen, A. Foerster, R. Schmidt: Phys. Rev. B **61**, 8276 (2000)
14. J.M. Moison, C. Guille, F. Houzay, F. Barthe, M. Van Rompay: Phys. Rev. B. **40**, 6149 (1989)
15. J.-M. Gerard, G. Le Roux: Appl. Phys. Lett. **62**, 3452 (1993)
16. O.M. Khreis, W.P. Gillin, K.P. Homewood: Phys. Rev. B **55**, 15813 (1997)

17. J.S. Tsang, C.P. Lee, S.H. Lee, K.L. Tsai, H.R. Chen: J. Appl. Phys. **77**, 4302 (1995)
18. S.S. Rao, W.P. Gillin, K.P. Homewood: Phys. Rev. B **50**, 8071 (1994)
19. R. Leon, D.R.M. Williams, J. Krueger, E.R. Weber, M.R. Melloch: Phys. Rev. B **56**, R4336 (1997)
20. A. Rosenauer, U. Fischer, D. Gerthsen, A. Förster: Appl. Phys. Lett. **71**, 3868 (1997)
21. P.B. Joyce, T.J. Krzyzewski, G.R. Bell, B.A. Joyce, T.S. Jones: Phys. Rev. B **58**, R15981 (1998)
22. I. Kegel, T.H. Metzger, A. Lorke, J. Peisl, J. Stangl, G. Bauer, J.M. Garcia, P.M. Petroff: Phys. Rev. Lett. **85**, 1694 (2000)

9 Electron Holography: AlAs/GaAs Superlattices

Although the CELFA technique described in Chap. 5 is readily applicable to ternary semiconductors such as $In_xGa_{1-x}As$ and $Cd_xZn_{1-x}Se$, the $Al_xGa_{1-x}As$ system is more difficult to analyze. It was shown in Sect. 5.3.5 that the compositional analysis of $In_xGa_{1-x}As$ and $Cd_xZn_{1-x}Se$ does not require a precise knowledge of the specimen thickness. In a thickness range of up to more than 40 nm, the assumption of a thickness of 15 nm is sufficient to obtain accurate results. However, a precise knowledge of the specimen thickness is crucial for $Al_xGa_{1-x}As$, as becomes obvious from Fig. 5.22c. The aim of this chapter is to demonstrate that electron holography is an elegant solution to this problem because it allows a precise evaluation of the local specimen thickness and an assessment of the local composition from only one hologram.

The procedure described here is based on electron holography and the exploitation of the chemically sensitive (002) reflection. We apply an off-axis imaging condition, with the (002) beam strongly excited and centered on the optical axis. The first sideband of the hologram is centered using an "empty" reference hologram obtained with a hole in the specimen. From the centered sideband, we use the phase of the central (000) reflection, and the amplitude of the (002) reflection to evaluate the local composition and the local specimen thickness in an iterative and self-consistent way. Delocalization effects that lead to a shift of the spatial information of the (000) and (002) reflections are taken into account. The application of the procedure is demonstrated with an AlAs/GaAs(001) SL with a period of 5 nm. The concentration profiles obtained are discussed in relation to segregation. The measured segregation efficiency is $R = 0.51 \pm 0.02$.

9.1 Experimental Setup

The epitaxial growth of the samples was carried out by G. Böhm at the Walter Schottky Institut, Technische Universität München, Garching. The layers were grown pseudomorphically by MBE on GaAs(001) substrates. The samples contained two AlAs/GaAs SLs and an InGaAs layer in between, separated by GaAs spacers of 30 nm thickness. In this chapter, we focus on the investigation of the AlAs/GaAs SL close to the sample surface, which

was capped with 10 nm GaAs. The nominal thickness of each of the AlAs
and GaAs layers was 2.5 nm. The substrate temperature was 480°C during
the growth of the SL considered here. The growth rate was 0.1 nm s^{-1} for
AlAs and 0.31 nm s^{-1} for GaAs, and the As$_4$ BEP was 0.93×10^{-5} Torr.

The TEM cross-section samples were prepared in the conventional way by
Ar$^+$ ion milling at an energy of 3 keV in a liquid-nitrogen-cooled specimen
holder. A Philips CM30 FEG/UT transmission electron microscope equipped
with an electron biprism, with a spherical-aberration constant $C_S = 0.65$ mm
and a point-to-point resolution of 0.18 nm was used. The holograms were
recorded with an on-line CCD camera fixed at the exit of a Gatan imaging
filter, with 1024×1024 picture elements (pixels).

9.2 Description of the Method

9.2.1 Imaging Condition

Our procedure is based on the exploitation of the phase $P(\mathbf{g}_{000})$ and the
modulus $A(\mathbf{g}_{002})$ of the (000) and (002) reflections of the centered sideband.
By using the sideband reflections (instead of those of the autocorrelation),
we gain the following advantages:

1. There is no effect of inelastic scattering, because the sideband is zero-loss
   filtered [1].
2. Owing to the linear image formation in the sideband, $A(\mathbf{g}_{002})$ is given by

$$A(\mathbf{g}_{002}) \propto a(\mathbf{g}_{002})a_{\text{ref.}} , \qquad (9.1)$$

   where $a_{\text{ref.}}$ is the (to a very good approximation) spatially homogeneous
   amplitude of the reference beam.
3. The phase $P(\mathbf{g}_{000})$ allows the measurement of the local specimen thickness t. This phase depends mainly on t, but is also influenced by the
   chemical composition to "second order".

The imaging condition applied is described in Sect. 5.1.2. The specimen
was tilted by about 5° from the [100] ZA orientation along the (004) Kikuchi
band, corresponding to a rotation around an axis perpendicular to the projected interface. The advantage compared with the ZA orientation is a larger
extinction distance [2] and an almost linear dependence of $P(\mathbf{g}_{000})$ on the
specimen thickness. The (002) reflection was centered on the optical axis to
minimize the effects of objective lens aberrations. The last free parameter
of the sample orientation is a rotation around an axis parallel to the projected interface, which defines the excitation error of the (002) reflection.
The following requirements must be satisfied: the intensity of the (002) beam
should be as great as possible in order to maximize the signal-to-noise ratio,
but on the other hand, small mistilts should have only a small effect on the

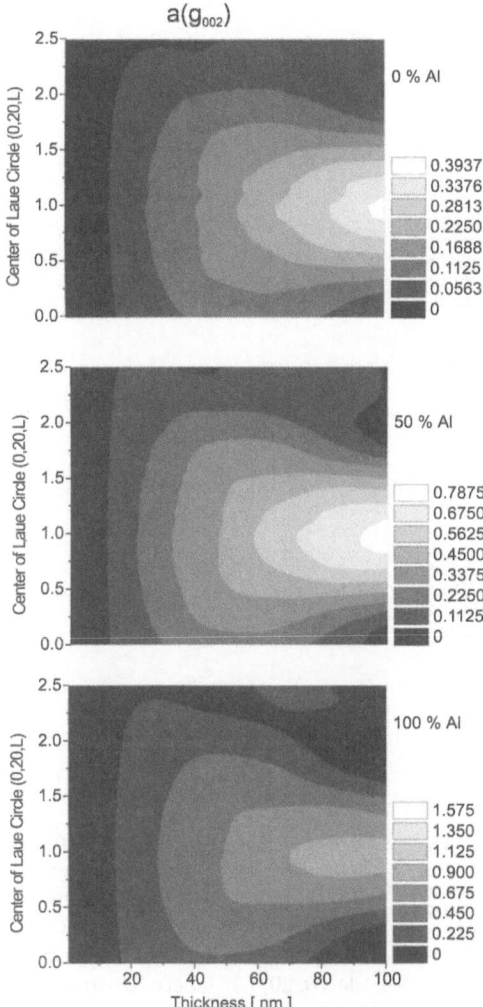

Fig. 9.1. Grayscale-coded map of the amplitude of the (002) beam (calculated for $Al_xGa_{1-x}As$ with $x = 0$, 50, 100% and an acceleration voltage of 300 kV) plotted versus the specimen thickness and versus the position of the Laue circle $(0, 20, L)$, where L varies between 0 and 2.5. Note that $L = 1.0$ corresponds to a strongly excited (002) beam without excitation error, whereas the excitation error of the (004) reflection vanishes for $L = 2.0$. The values in the legends refer to the lower boundaries of the corresponding intervals

evaluation. Figure 9.1 clearly shows, for the example of $Al_xGa_{1-x}As$ with Al concentrations of 0, 50 and 100%, that both conditions are well fulfilled for a strongly excited (002) beam because this occurs at a maximum of $A(\mathbf{g}_{002})$, corresponding to the center of the Laue circle being at $(0, 20, L=1)$. Figures 9.1 and 9.2 were obtained from Bloch wave calculations with the EMS program package [3]. Figure 9.2 reveals that the mistilt sensitivity of $P(\mathbf{g}_{000})$ is sufficiently small for a strongly excited (002) beam, which would not be the case for an imaging condition where the excitation error vanishes for the (004) beam ($L = 2$ in Fig. 9.2).

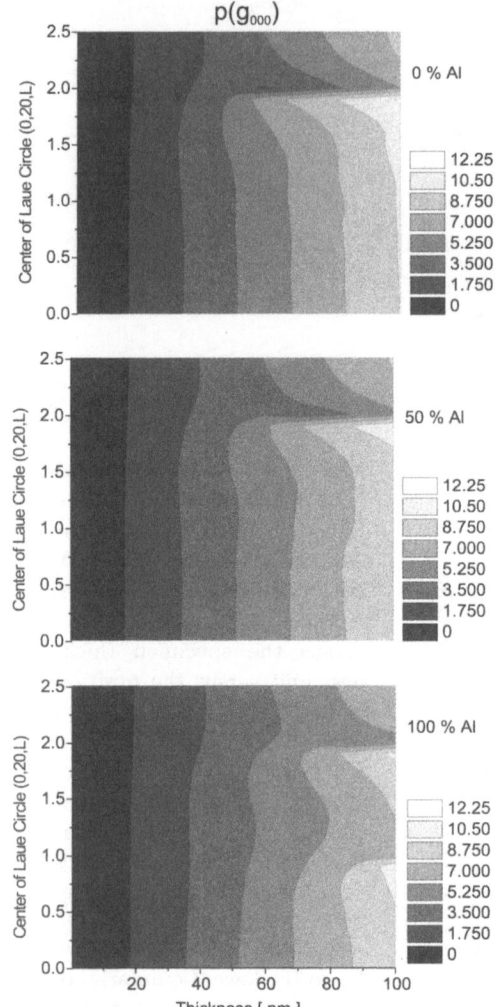

Fig. 9.2. Grayscale-coded map of the phase of the (000) beam plotted versus the specimen thickness and the position of the Laue circle $(0, 20, L)$, where L varies between 0 and 2.5

9.2.2 Experimental Details

In practice, the specimen orientation described above was achieved by an initial rough alignment by selected-area diffraction and a subsequent fine adjustment by maximizing the intensity of the bright AlAs layers in the image. A hologram of the AlAs layer and an "empty" reference hologram corresponding to a hole in the specimen were then taken under elliptical illumination [4].

The sideband of the hologram was centered by the following procedure [5]. The reference hologram was Fourier-transformed and a circular area with center $\mathbf{G}_0$ and radius G was selected with a mask around the appropriate sideband. Each pixel $\mathbf{g}$ of the Fourier-transformed image was then multiplied with

Fig. 9.3. (a) Hologram of the AlAs/GaAs SL. (b) Extracted phase $P(\mathbf{g}_{000})$ of the central beam and (c) extracted amplitude $A(\mathbf{g}_{002})$ of the (002) reflection of the centered sideband

by filter $\exp(-(|\mathbf{g} - \mathbf{G}_0|/G)^8)$. The result was inversely Fourier-transformed, yielding a reference image

$$R(\mathbf{r}) = \rho(\mathbf{r})\exp(i\Phi_R(\mathbf{r})) , \qquad (9.2)$$

where $\mathbf{r} = (x, y)$ is the coordinate vector in the image and $\rho(\mathbf{r})$ is the modulus of $R(\mathbf{r})$. The hologram of the specimen $H(\mathbf{r})$ was modified according to

$$H_c(\mathbf{r}) = H(\mathbf{r})\exp(-i\Phi_R(\mathbf{r})) . \qquad (9.3)$$

In the resulting hologram H_c, the appropriate sideband was centered with subpixel accuracy. A further advantage is that this procedure leads to an automatic correction of geometric distortions such as anisotropic magnification [6] and artificial bending of the hologram fringes [6], as well as image distortions caused by the Gatan imaging filter. In the next step, local maps of $P(\mathbf{g}_{000})$ and $A(\mathbf{g}_{002})$ were obtained by calculating the Fourier transform of H_c and filtering circular areas around the central reflection and the (002) reflection of the centered sideband as described above. The phase $P(\mathbf{g}_{000})$ that is obtained after the inverse Fourier transform has a range of values $P(\mathbf{g}_{000}) \in [0, 2\pi[$. To get rid of the phase jumps of 2π that occurred, the phase $P(\mathbf{g}_{000})$ was "unwrapped" [7] for further evaluation. Figure 9.3a shows the initial hologram H and Figs. 9.3b,c the extracted maps for $P(\mathbf{g}_{000})$ and $A(\mathbf{g}_{002})$, respectively.

For the subsequent steps of the evaluation procedure, the image was subdivided into square image unit cells by the formation of a regularly spaced grid with two sets of perpendicular grid lines. The grid lines of one of the sets ran parallel to the interfaces. The spacing of grid lines corresponded to 1 ML (0.28 nm). The dimensions and orientation of the grid were computed from the position of the (002) reflection in the centered sideband. The grid lines of each of the two sets were numbered sequentially so that each cell could be labeled as $C(i, j)$. We assigned the variables $A(\mathbf{g}_{002}, i, j)$, $P(\mathbf{g}_{000}, i, j)$, $t(i, j)$ (corresponding to the specimen thickness) and $x(i, j)$ (corresponding to the concentration) to each of the unit cells. The values of $A(\mathbf{g}_{002}, i, j)$ were obtained by averaging the calculated values of $A(\mathbf{g}_{002})$ (shown in Fig. 9.3c) inside the area of each unit cell $C(i, j)$, and an analogous procedure was applied to obtain $P(\mathbf{g}_{000}, i, j)$. The following sections describe the computation of the values of $t(i, j)$ and $x(i, j)$.

Figure 9.4 shows $A(\mathbf{g}_{002}, i, j)$ and $P(\mathbf{g}_{000}, i, j)$ averaged along the grid lines parallel to the interfaces inside the region indicated by the black frame in Fig. 9.3a. The vacuum region is at the right-hand side in Fig. 9.4. The phase $P(\mathbf{g}_{000})$ jumps by about 4.5 rad at the vacuum–specimen transition and oscillates in the region of the AlAs/GaAs SL. Stronger oscillations are observed in $A(\mathbf{g}_{002})$. The minima of $P(\mathbf{g}_{000})$ and the maxima of $A(\mathbf{g}_{002})$ correspond to AlAs. Note that all of the graphs of $A(\mathbf{g}_{002})$, $P(\mathbf{g}_{000})$, t and x that will be shown in the following were obtained by averaging in the same way as in Fig. 9.4.

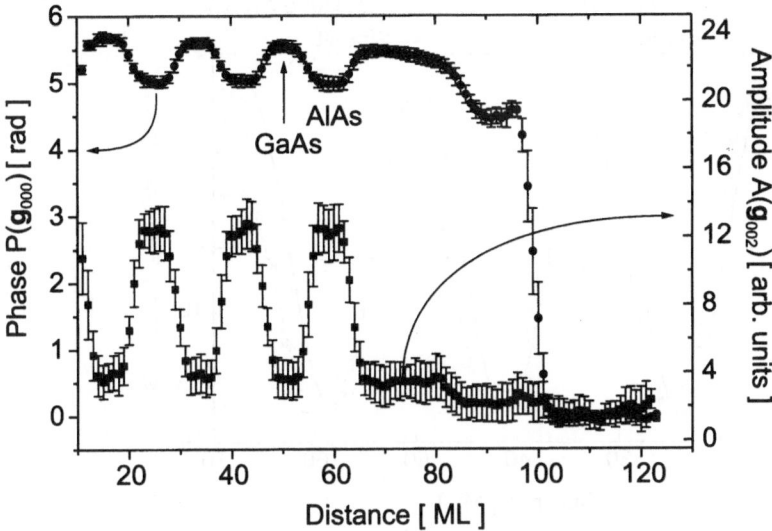

Fig. 9.4. Phase $P(\mathbf{g}_{000})$ and amplitude $A(\mathbf{g}_{002})$ plotted versus the distance in the growth direction. The unit "1 ML" corresponds to 0.28 nm. Each data point was obtained by averaging along one grid line (parallel to the interfaces) in the region indicated in Fig. 9.3a. The *error bars* correspond to $\pm\sigma$, where σ is the standard deviation

9.2.3 Delocalization

Figure 9.5 shows the same data as in Fig. 9.4 but restricted to the SL region. Obviously, the oscillations of $P(\mathbf{g}_{000})$ and $A(\mathbf{g}_{002})$ are shifted relative to each other. This effect is due to delocalization [8, 9], which is, according to Sect. 3.2.4, given by

$$D = |C_{\mathrm{S}}\lambda^3\mathbf{g}^3 + \lambda\,\Delta f\,\mathbf{g}| \, . \tag{9.4}$$

In the image plane, the spatial distributions of the information contained in the Fourier components of the image wave function that correspond to the undiffracted beam and to the (002) beam are shifted relative to each other. While $D = 0$ for the (002) beam, which runs parallel to the optical axis, a displacement of the information occurs for the undiffracted beam, which is tilted. This effect could be avoided by multiplying the reconstructed sideband by an appropriate phase plate [6]. However, for the present application it is sufficient to measure the effect of the delocalization and realign $A(\mathbf{g}_{002})$ towards $P(\mathbf{g}_{000})$. Fitting appropriate sine curves to the oscillations of $A(\mathbf{g}_{002})$ and $P(\mathbf{g}_{000})$ (see Fig. 9.5) revealed a shift of 1.1 ML. To take this effect into account, each of the values $A(\mathbf{g}_{002}, i, j)$ associated with a unit cell $C(i, j)$ was recomputed by averaging $A(\mathbf{g}_{002})$ shown in Fig. 9.3c over a region that corresponded to the size of the unit cell but was shifted by 1.1 ML. As a

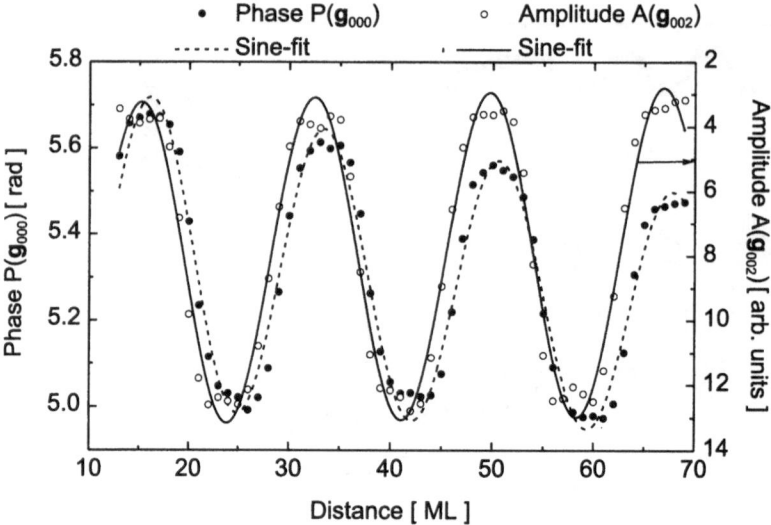

Fig. 9.5. Phase $P(\mathbf{g}_{000})$ and amplitude $A(\mathbf{g}_{002})$ plotted versus the distance in the growth direction. The *solid* and *dashed lines* are sine fits used to obtain the shift between $P(\mathbf{g}_{000})$ and $A(\mathbf{g}_{002})$

further result, the sine fit yields an SL period of $(17.2\pm0.1)\,\text{ML} = (4.86\pm0.03)$ nm.

9.2.4 Iterative Computation of the Composition

The evaluation of the composition was based on tabulated values of $a(\mathbf{g}_{002}, x, t)$ and $p(\mathbf{g}_{000}, x, t)$, where the Al concentration x varied between 0 and 100% (step size 1%) and the specimen thickness t between 0.5 and 150 nm (step size 0.5 nm). The Bloch wave calculations were performed with the EMS program package [3], using the electron scattering factors calculated by Doyle and Turner [10]. The EMS programs "bul" and "bz2" with the option "t" yield a mean inner potential of 15.27 V for GaAs and 13.98 V for AlAs. In [11], the scattering factors of Doyle and Turner were also used, and a mean inner potential of 15.19 V was obtained for GaAs. A recent experimental study using electron holography [12] yielded a value of (14.53 ± 0.17) V. The application of the EMS program "as is", therefore, results in an evaluated thickness that is about 5% too small. However, we shall show that this deviation is of minor importance in the evaluation of the concentration. Figure 9.6 gives the result of the Bloch wave calculation, showing the normalized modulus $a(\mathbf{g}_{002}, x, t)/a(\mathbf{g}_{002}, 0, t)$ and the phase $p(\mathbf{g}_{000}, x, t)$ as a function of the specimen thickness t and the Al concentration x. Below $t = 80$ nm, $p(\mathbf{g}_{000}, x, t)$ depends almost linearly on the specimen thickness. The dependence on the Al concentration is weaker and, therefore, can be regarded as a "second-order" effect, which is also reflected in the experimental data for

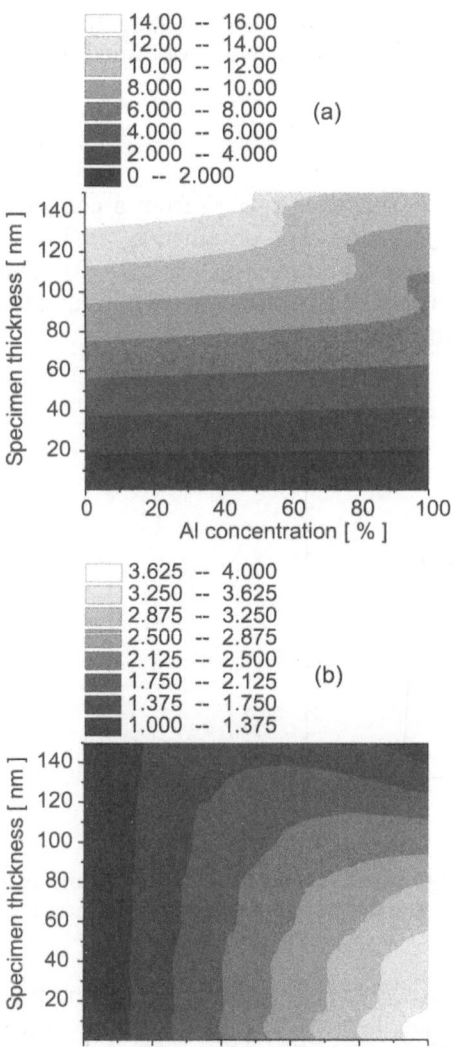

14.00 -- 16.00
12.00 -- 14.00
10.00 -- 12.00
8.000 -- 10.00
6.000 -- 8.000   (a)
4.000 -- 6.000
2.000 -- 4.000
0 -- 2.000

3.625 -- 4.000
3.250 -- 3.625
2.875 -- 3.250
2.500 -- 2.875
2.125 -- 2.500
1.750 -- 2.125   (b)
1.375 -- 1.750
1.000 -- 1.375

Fig. 9.6. Contour plots of (a) the phase $p(\mathbf{g}_{000})$ and (b) the normalized amplitude $a(\mathbf{g}_{002}, x, t)/a(\mathbf{g}_{002}, 0, t)$ versus the specimen thickness and the Al concentration. The calculation was performed with the EMS program package for an acceleration voltage of 300 kV. The center of the Laue circle was set to $(0, 20, 1)$, corresponding to a strongly excited (002) beam

$P(\mathbf{g}_{000})$ shown in Fig. 9.4. Figure 9.6b shows the composition dependence of the modulus of the (002) beam, which is unambiguous for $t < 100$ nm. Figure 9.6b also reveals that an error in the thickness measurement of 5% has an effect on the Al concentration obtained that is sufficiently small to be neglected. The Al concentration was evaluated by the following procedure.

In the first iteration, we estimated the local specimen thickness $t_1(i,j)$ for each image unit cell $C(i,j)$ by a comparison of $P(\mathbf{g}_{000}, i, j)$ with the tabulated values of $p(\mathbf{g}_{000}, x = 0, t)$. This is a good first approximation because the composition dependence is a "second-order" effect, as we showed above.

Compositions $x_1(i,j)$ were then estimated by comparing $\alpha A(\mathbf{g}_{002}, i, j)$ with the tabulated values of $a(\mathbf{g}_{002}, x, t_1(i,j))$. The factor α was obtained from a GaAs region using

$$a(\mathbf{g}_{002}, 0, t_1(i,j)) = \alpha A(\mathbf{g}_{002}, i, j) . \tag{9.5}$$

For each subsequent iteration $n > 1$, we computed $t_n(i,j)$ from a comparison of $P(\mathbf{g}_{000}, i, j)$ with $p(\mathbf{g}_{000}, x_{n-1}(i,j), t)$ and then obtained $x_n(i,j)$ from minimizing the difference between $\alpha A(\mathbf{g}_{002}, i, j)$ and $a(\mathbf{g}_{002}, x, t_n(i,j))$. The procedure was halted (at $n = 4$) as soon as stable solutions were obtained for the specimen thickness and the Al concentration.

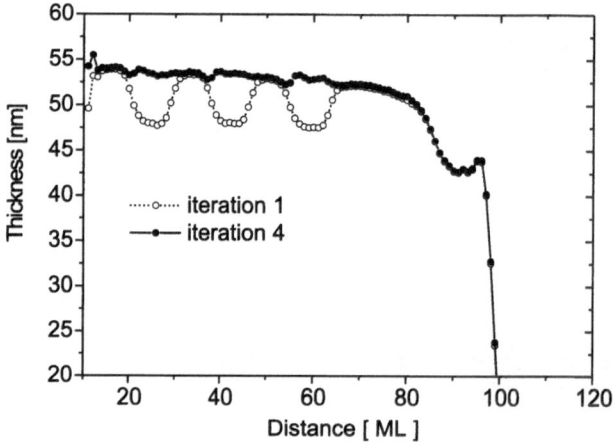

Fig. 9.7. Averaged specimen thickness obtained in the first and fourth iterations of the evaluation procedure, plotted versus the distance in the growth direction

Averaged values of $t_{1,4}(i,j)$ and $x_{1,4}(i,j)$ are shown in Figs. 9.7 and 9.8, respectively. Obviously, the "artificial" oscillations of $t_1(i,j)$ decrease with increasing n, and the evaluated specimen thickness approaches a smooth profile. The horizontal lines in Fig. 9.8 indicate that the difference between $x_1(i,j)$ and $x_4(i,j)$ increases with increasing Al concentration. Note that the averaged Al concentration is smaller than 100% in the AlAs regions and larger than 0% in the GaAs. This is not a systematic error, but is caused by noise combined with the fact that the evaluated local Al concentration can only take values between 0 and 100%.

To give an impression of the local accuracy of the method, grayscale-coded maps of $t_1(i,j)$, $t_4(i,j)$ and $x_4(i,j)$ are shown in Fig. 9.9.

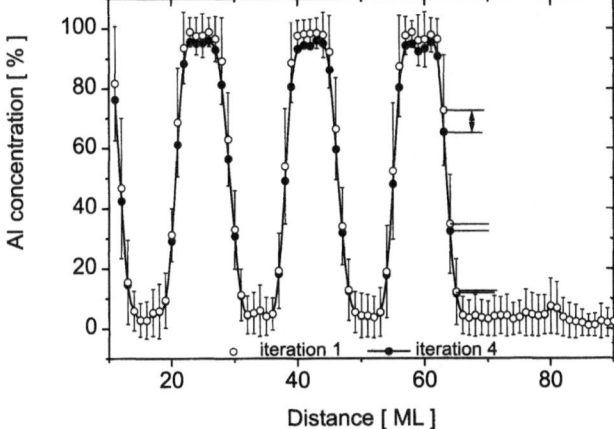

Fig. 9.8. Averaged composition profiles obtained in the first and fourth iterations of the evaluation procedure, plotted versus the distance in the growth direction. The *error bars* correspond to $\pm\sigma$, where σ is the standard deviation

9.3 Discussion

Figure 9.10 displays the evaluated concentration profile. The error bars correspond to $\pm\sigma$, where σ is the standard deviation obtained by averaging along the (002) fringes. In [13, 14], it is shown that segregation of Ga plays an important role in the formation of an AlAs/GaAs heterointerface during growth. We have compared our measurements with two models of segregation.

Firstly, we used McLean's formula. From (6.5), we obtain

$$\ln\left(\frac{x_{\text{bulk}}}{1 - x_{\text{bulk}}}\right) = \frac{E_{\text{S}}}{kT} + \ln\left(\frac{x_{\text{surf}}}{1 - x_{\text{surf}}}\right) , \tag{9.6}$$

where x_{surf} and x_{bulk} denote the surface and bulk compositions of Al, and E_{S} denotes the difference between the bulk and surface free energies [13]. For the calculation of the concentration profile, we used $E_{\text{S}} = 0.1$ eV, given in [13]. Note that the segregation energy E_{S} is positive here, whereas it was negative in Sect. 8.3.4. This is simply related to the fact that the concentration of the segregating element Ga in $\text{Al}_x\text{Ga}_{1-x}\text{As}$ is denoted by $1 - x$, whereas the segregating element In in $\text{In}_x\text{Ga}_{1-x}\text{As}$ has a concentration denoted by x. We used a substrate temperature of $T = 753$ K and assumed [13] that an equilibrium state was reached during the deposition of every ML, in accordance with the $x_{\text{surf}}(x_{\text{bulk}})$ relation (6.5) and the matter conservation relation. The underlying ML was then "frozen" for the rest of the growth process. Using $x_{\text{surf}}^{(1)} + x_{\text{bulk}}^{(1)} = 0.5$ for the first ML of the AlAs layer, $x_{\text{surf}}^{(n)} + x_{\text{bulk}}^{(n)} = x_{\text{surf}}^{(n-1)} + 1$ for $2 \leq n \leq 9$ and $x_{\text{surf}}^{(n)} + x_{\text{bulk}}^{(n)} = x_{\text{surf}}^{(n-1)}$ for $n > 9$, we obtained a profile corresponding to a nominally 8.5 ML thick AlAs layer. The result is shown as a dotted curve in Fig. 9.10.

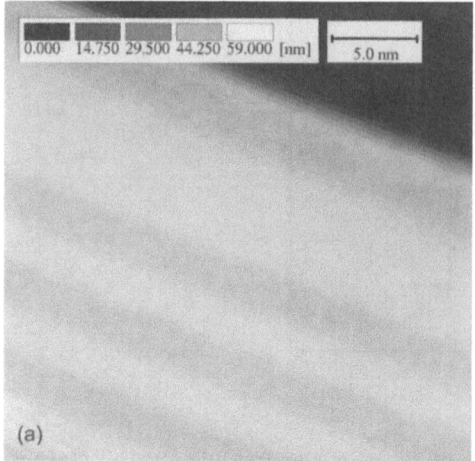

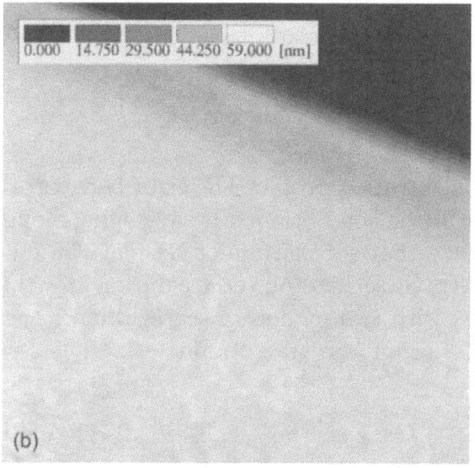

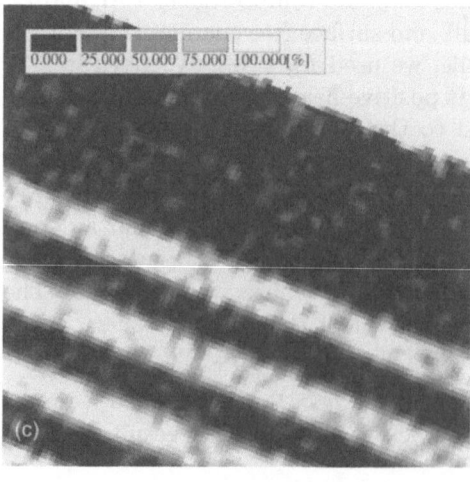

Fig. 9.9. Grayscale-coded maps (**a**) of the evaluated thickness after the first iteration and (**b**) after the fourth iteration. (**c**) Map of the Al concentration (fourth iteration)

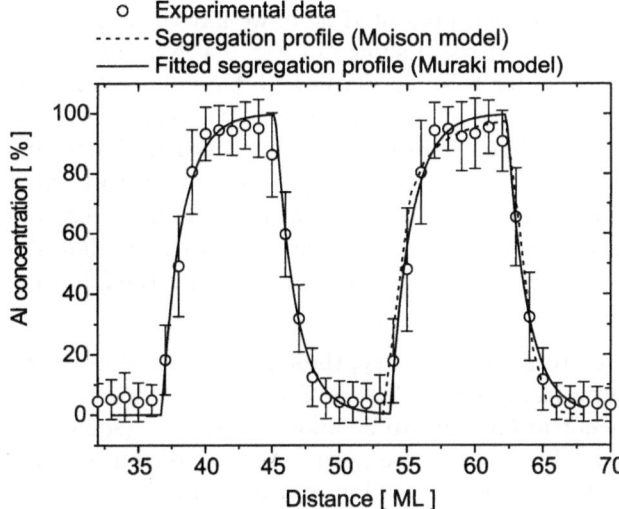

Fig. 9.10. Averaged experimental and calculated composition profiles plotted versus the distance in the growth direction. The error bars correspond to $\pm\sigma$, where σ is the standard deviation

Secondly, we used the phenomenological description suggested by Muraki et al. [15] in accordance with (6.6):

$$x(n) = \begin{cases} 0 & : \quad n \le 0 \\ (1 - R^n) & : \quad 0 < n \le N \ , \\ (1 - R^N)R^{n-N} & : \quad n > N \end{cases} \tag{9.7}$$

where R is the segregation probability and N the number of deposited ML of AlAs. Assuming a two-layer profile $x_{2L}(n) = x(n - n_1) + x(n - n_2)$ and using R and N as fit parameters, (9.7) was fitted to the experimental profile shown in Fig. 9.10. We obtained $R = 0.51 \pm 0.02$ and $N = (8.52 \pm 0.09)$ ML. The SL period is $n_2 - n_1 = (17.1 \pm 0.2)$ ML, in agreement with the value obtained from the sine fit in Fig. 9.5. The corresponding profile is displayed as a solid curve in Fig. 9.10.

Obviously, both calculated profiles describe the measured profile within the tolerance of the error bars. Note that the profile obtained from the McLean formula is not a fit but is based on a value of E_S obtained by electron spectroscopy experiments [13]. We do not find a significant deviation of the profile obtained with McLean's equation from the experimental profile, as was reported in [14].

The quantitative chemical-lattice-imaging method presented by Ourmazd et al. [16] also relies on the presence of chemical reflections. The sensitivity of that method was described by the standard deviation σ of the projected image unit cell vectors from their respective template vectors. For ion-milled specimens, the projected template vectors for GaAs and $Al_{0.4}Ga_{0.6}As$ were separated by 5σ [16]. For the method presented here, we have estimated the sensitivity from the error bars of $A(\mathbf{g}_{002})$ shown in Fig. 9.4 in an analogous way, and obtain a separation of 9σ between GaAs and AlAs. Therefore, the sensitivity obtained in [16] is better by a factor of 1.4. We expect that the sensitivity of the procedure outlined here could be improved by increasing the exposure time, which was limited to 2 s owing to specimen drift. Note that the stringent coherence requirement in the plane of the biprism leads to a small electron beam current density [17]. Compared with the method described in Ref. [16], the approach using electron holography has the following advantages:

1. In [16], two template vectors are necessary; these are extracted from two regions of known composition, e.g. a GaAs and an $Al_{0.4}Ga_{0.6}As$ region. Therefore, the Al concentration x has to be known a priori in (some part of) the Al_xGa_xAs layer. In contrast, we need only one reference region (e.g. the GaAs substrate) for the calculation of α in (9.5).
2. Our procedure is not restricted to certain intervals of specimen thickness and defocus.

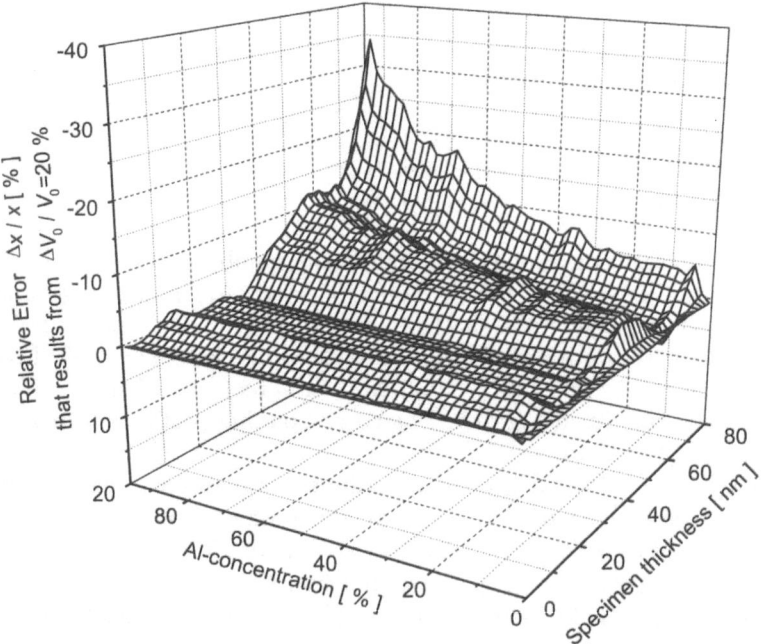

Fig. 9.11. Relative error $\Delta x/x$ plotted versus the Al concentration x and the specimen thickness t corresponding to $\Delta V_0/V_0 = 20\%$, calculated from (9.8)

Note that the evaluated Al concentration is rather tolerant with respect to errors $\Delta V_0/V_0$ in the mean inner potential V_0. In the following, we estimate the error $\Delta x/x$ caused by a fictitious, large uncertainty of $\Delta V_0/V_0 = 20\%$. Figure 9.6a shows that $P(\mathbf{g}_{000}) \propto t$ is a good approximation for a specimen thickness below 80 nm. Using $P(\mathbf{g}_{000}) \propto V_0$ we obtain the error in the measured specimen thickness as $\Delta t/t \approx \Delta V_0/V_0$. The error in the composition measurement can then be estimated from

$$
\frac{\Delta x}{x} \approx \frac{1}{x} \left(\frac{\partial a(\mathbf{g}_{002}, x, t)}{\partial x} \right)^{-1}
$$
$$
\times \left[\frac{a(\mathbf{g}_{002}, x, t+t/10)}{a(\mathbf{g}_{002}, 0, t+t/10)} - \frac{a(\mathbf{g}_{002}, x, t-t/10)}{a(\mathbf{g}_{002}, 0, t-t/10)} \right]. \tag{9.8}
$$

The result is shown in Fig. 9.11. Obviously, the error induced by uncertainties in the mean inner potential is rather small ($\Delta x/x < 12\%$) for a specimen thickness below 70 nm. For the hologram evaluated here, we estimate an error $\Delta V_0/V_0 \approx 5\%$, which induces an error in the composition evaluation of $\Delta x/x < 3\%$.

9.4 Conclusion

In this chapter we have suggested a method for compositional analysis that uses applying electron holography. Electron holography offers the advantage of zero-loss filtered data and linear image formation in the sideband, combined with the possibility of phase detection. With the reconstructed sideband, we can exploit the phase of the central beam and the modulus of a chemically sensitive beam. Although these two channels of information depend not only on the composition but also on the specimen thickness, it was possible to obtain an unambiguous solution by use of a simple iterative algorithm. The application of this technique was demonstrated with an AlAs/GaAs SL, where we obtained concentration profiles that were in good agreement with concentration profiles predicted by Moison et al. [13] on the basis of electron spectroscopy experiments. In the context of the description of segregation by Muraki et al. [15], we obtained a segregation probability $R = 0.51 \pm 0.02$. The technique described here is applicable to a wide variety of systems, for example $In_x Ga_{1-x} As$ and $Cd_x Zn_{1-x} Se$, provided the signal-to-noise ratio of the reconstructed chemically sensitive beam is sufficiently large.

References

1. D. Van Dyck, H. Lichte, J.C.H. Spence: Ultramicroscopy **81**, 187 (2000)
2. C.L. Jia, A. Thust, G. Jacob, K. Urban: Ultramicroscopy **49**, 330 (1993)
3. P.A. Stadelmann: Ultramicroscopy **51**, 131 (1987)

4. W.-D. Rau: *Ein on-line Bildverarbeitungssystem für die Bildebenen-Off-Axis Holographie mit Elektronen*. Ph.D. Thesis, Eberhard-Karls-Universität, Tübingen (1994)
5. M. Lehmann: *Numerische Rekonstruktion der aberrationsfreien Objektwelle aus off-axis Elektronenhologrammen*. Ph.D. Thesis, Eberhard-Karls-Universität, Tübingen (1997)
6. H. Lichte, D. Geiger, A. Harscher, E. Heindl, M. Lehmann, D. Malamidis, A. Orchowski, W.-D. Rau: Ultramicroscopy **64**, 67 (1996)
7. T.R. Judge, P.J. Bryanston-Cross: Opt. Lasers Eng. **21**, 199 (1994)
8. H. Lichte: Ultramicroscopy **38**, 13 (1991)
9. A. Thust, W.M.J. Coene, M. Op de Beeck, D. Van Dyck: Ultramicroscopy **64**, 211 (1996)
10. P.A. Doyle, P.S. Turner: Acta Cryst. A **24**, 390 (1968)
11. D. Rez, P. Rez: Acta Cryst. A **50**, 481 (1994)
12. M. Gajdardziska-Josifovska, M.R. McCarnez, W.J. de Ruijter, D.J. Smith, J.K. Weiss, J.M. Zuo: Ultramicroscopy **50**, 285 (1993)
13. J.M. Moison, C. Guille, F. Houzay, F. Barthe, M. Van Rompay: Phys. Rev. B **40**, 6149 (1989)
14. W. Braun, A. Trampert, L. Däweritz, K.H. Ploog: Phys. Rev. B **55**, 1689 (1997)
15. K. Muraki, S. Fukatsu, Y. Shirakia, R. Ito: Appl. Phys. Lett. **61**, 557 (1992)
16. A. Ourmazd, F.H. Baumann, M. Bode, Y. Kim: Ultramicroscopy **34**, 237 (1990)
17. B.M. Mertens, M.H.F. Overwijk, P. Kruit: Ultramicroscopy **77**, 1 (1999)

10 Outlook

In this book, methods were described for the compositional analysis of ternary sphalerite-type semiconductor nanostructures, based upon high-resolution transmission electron microscopy. These methods were applied to investigate the structure and chemical composition of quantum wells and quantum dots, which are an important topic in solid-state physics research at present.

Molecular-beam epitaxy and metal-organic vapor phase epitaxy are the most established techniques for the growth of semiconductor nanostructures. These methods allow control of the deposition with the highest accuracy that can be achieved at present. However, the experimental results obtained demonstrate that the growth of nanostructures is governed by processes such as segregation and atomic kinetics on the growth surface, which are not yet well understood. These novel insights, described in Part III, were only made possible by the methods outlined in this book in Part II, which allowed measurement of the chemical composition of ternary $In_xGa_{1-x}As$ and $Cd_xZn_{1-x}Se$ nanostructures with a previously unachieved combination of high accuracy and almost atomic-scale spatial resolution.

The methods for structural analysis with atomic-scale spatial resolution that were described in this book are of general importance. First, an understanding of the optical properties of semiconductor nanostructures requires detailed structural data on an atomic scale. Second, the optimization of the performance of optoelectronic devices is based upon precise control of the sizes and compositions of the structures in those devices, which requires an accurate knowledge of all processes that occur during the epitaxial growth. However, the investigation of InGaAs Stranski–Krastanov layers revealed significant deviations between the putative and the real morphologies, which indicates the existence of epitaxial processes that are not fully understood at present. The insufficient knowledge of the growth mechanisms can be explained by a lack of characterization methods that would allow a sufficiently accurate analysis of the chemical composition with an atomic-scale spatial resolution. This gap successfully has been closed for a large variety of sphalerite-type material systems by the methods outlined in Part II of the present book, as becomes obvious from the applications described in Part III and from the large number of related publications cited in Chap. 1.

Although the CELFA method turned out to be extremely useful for the evaluation of composition profiles, one should keep in mind that it is based on a comparison with Bloch wave simulations. These are presently carried out with the EMS program package, using the Doyle and Turner atomic scattering factors in the free-atom approximation. In this approximation, the crystal is treated as a 3D array of free, noninteracting atoms, and the redistribution of charge in between the atoms due to the atomic bonding is not taken into account. Another way to obtain electron-scattering structure factors that should be explored in future is the ab initio calculation of the local crystal potential. Density functional theory offers one possibility for such computations. One may expect that the free-atom approximation combined with the CELFA method will yield compositions that need correction with a (constant) factor, which most probably can be obtained from density functional theory. A second problem that now concerns composition evaluation with electron holography is the absence of accurate theoretical or experimental data for the mean inner potential of most of the III–V and II–VI semiconductors. In conclusion, the main goals that we are aiming at in the near future is the ab initio computation and experimental measurement of electron scattering factors, mean inner potentials and Debye–Waller factors of technologically interesting semiconductors to provide the necessary basis for an advanced quantitative analysis of elastic-electron-scattering data.

Appendices

A List of Acronyms Used

AFM	atomic-force microscopy
AOI	area of interest
BEP	beam-equivalent pressure
BF	bright-field
CA	correspondence analysis
CCD	charge-coupled device
CELFA	composition evaluation by lattice fringe analysis
CVD	chemical vapor deposition
DALI	digital analysis of lattice images
DF	dark-field
DFT	discrete Fourier transform
EDX	energy-dispersive X-ray analysis
EELS	electron energy loss spectroscopy
EFTEM	energy-filtered transmission electron microscopy
EMS	electron microscopy simulations [1]
FE	finite-element
FEG	field emission gun
FEM	finite-element method
FFT	fast Fourier transform
FOLZ	first-order Laue zone
FP	frank partial (dislocation)
FWHM	full width at half maximum
HML	half monolayer
HRTEM	high-resolution transmission electron microscopy
IDFT	inverse discrete Fourier transform
LFC	local Fourier coefficient
MAL	maximum-likelihood
MBE	molecular-beam epitaxy
ML	monolayer
MS	multislice
OA	off zone axis
PL	photoluminescence spectroscopy
QD	quantum dot

QUANTITEM	quantitative analysis of the information from transmission electron micrographs
SIMS	secondary-ion mass spectroscopy
SK	Stranski–Krastanov (growth mode)
SL	superlattice
SSL	short-period superlattice
TEM	transmission electron microscopy
WL	wetting layer (of an SK layer)
ZA	zone axis
ZOLZ	zero-order Laue zone

References

1. P.A. Stadelmann: Ultramicroscopy **51**, 131 (1987)

B Fourier Transform, Convolutions and δ-"Function"

The following expressions incorporate the sign conventions recommended as the standard conventions in [1, 2].

B.1 Fourier Transform

The Fourier transform from real to reciprocal space in two or more dimensions is given by

$$\widetilde{f}(\mathbf{k}) = \mathcal{F}f(\mathbf{r}) = \int\limits_{-\infty}^{\infty} f(\mathbf{r}) \exp\{2\pi i \mathbf{k} \cdot \mathbf{r}\} \, d\mathbf{r} \;, \tag{B.1}$$

where $\mathbf{r}$ is the coordinate in real space and $\mathbf{k}$ is related to reciprocal space. The inverse Fourier transform from reciprocal to real space is given by

$$f(\mathbf{r}) = \mathcal{F}^{-1}\widetilde{f}(\mathbf{k}) = \int\limits_{-\infty}^{\infty} \widetilde{f}(\mathbf{k}) \exp\{-2\pi i \mathbf{k} \cdot \mathbf{r}\} \, d\mathbf{k} \;. \tag{B.2}$$

We have the following results:

$$f(\mathbf{r}) = \mathcal{F}^{-1}\mathcal{F}f(\mathbf{r}) \;, \tag{B.3}$$

$$f(-\mathbf{r}) = \mathcal{F}\mathcal{F}f(\mathbf{r}) \;. \tag{B.4}$$

B.2 Convolution

The convolution integral of two functions $f(\mathbf{r})$ and $g(\mathbf{r})$ is given by

$$f(\mathbf{r}) \otimes g(\mathbf{r}) = \int\limits_{-\infty}^{\infty} f(\mathbf{r}')g(\mathbf{r} - \mathbf{r}') \, d\mathbf{r}' \;. \tag{B.5}$$

The multiplication and convolution theorems can be stated as:

$$\mathcal{F}[f(\mathbf{r})g(\mathbf{r})] = [\mathcal{F}f(\mathbf{r})] \otimes [\mathcal{F}g(\mathbf{r})] \;, \tag{B.6}$$

$$\mathcal{F}[f(\mathbf{r}) \otimes g(\mathbf{r})] = [\mathcal{F}f(\mathbf{r})][\mathcal{F}g(\mathbf{r})] \;. \tag{B.7}$$

B.3 δ-"Function"

The δ-distribution is a linear map defined by $\delta(f) := f(0)$, $\delta : \mathcal{D} \to \mathbb{R}$, where $\mathcal{D}$ is the vector space of test functions $f : \mathbf{R}^n \to \mathbf{R}$. The δ-distribution therefore has the following properties:

$$\int_{-\infty}^{\infty} \delta(\mathbf{r}) f(\mathbf{r}) \, d\mathbf{r} = f(0) , \tag{B.8}$$

$$\delta(\mathbf{r}/c) = |c|\delta(\mathbf{r}) , \tag{B.9}$$

$$f(\mathbf{r})\delta(\mathbf{r} - \mathbf{r}_0) = f(\mathbf{r}_0)\delta(\mathbf{r} - \mathbf{r}_0) , \tag{B.10}$$

$$\mathcal{F}\delta(\mathbf{r}) = 1 , \tag{B.11}$$

$$\mathcal{F}\delta(\mathbf{r} - \mathbf{r}_0) = \exp(2\pi i \mathbf{k} \cdot \mathbf{r}_0) , \tag{B.12}$$

$$\mathcal{F}^{-1}\delta(\mathbf{k} - \mathbf{k}_0) = \exp(-2\pi i \mathbf{k}_0 \cdot \mathbf{r}) . \tag{B.13}$$

References

1. *International Tables for Crystallography.* ed. by T. Hahn, Vol. B, 1st edn. (Kluwer Academic, Dordrecht, 1991)
2. P.R. Buseck, J.M. Cowley, L. Eyring: *High-Resolution Transmission Electron Microscopy*, 1st edn. (Oxford University Press, Oxford, 1988)

Color Plates

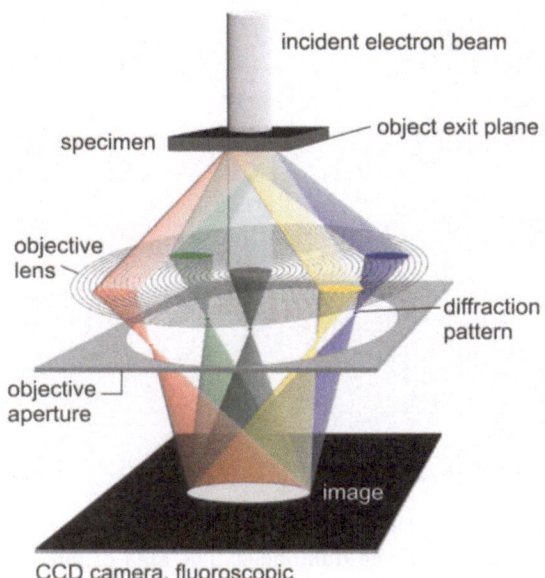

Plate 1. Schematic drawing showing the functional principle of the transmission electron microscope. Each diffracted beam is focused to a spot in the diffraction pattern, which occurs in the back focal plane of the objective lens. Beams selected with the objective aperture interfere in the image

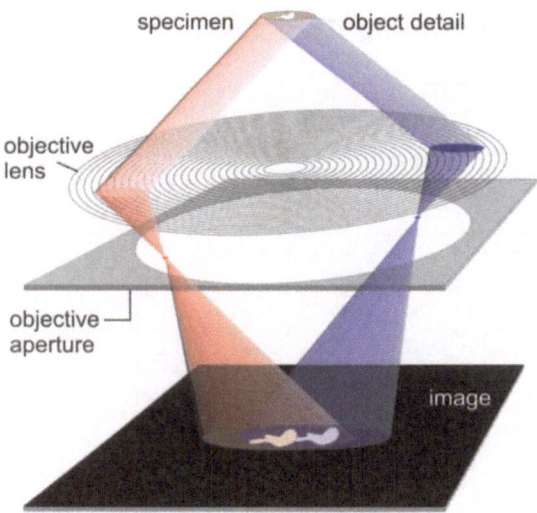

Plate 2. Schematic drawing explaining the effect of delocalization. At the top of the ray path, an object detail is shown. The image detail is transferred to the image plane by two diffracted beams. In the image plane, each beam generates an image of the object detail. Owing to delocalization, the two images of the object detail do not coincide, but are *delocalized*

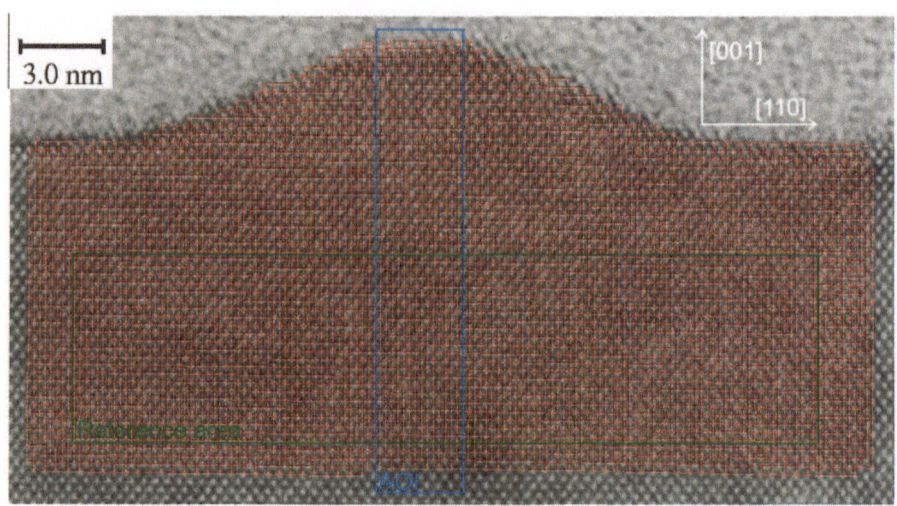

Plate 3. HRTEM image of an $In_xGa_{1-x}As/GaAs(001)$ SK island containing a grid that connects the local brightness maxima of the dumbbells. The area of interest (AOI) (*blue frame*) was used for the determination of the In concentration inside the island. The reference area (*green frame*) was used for the calculation of the basis vectors of the reference lattice

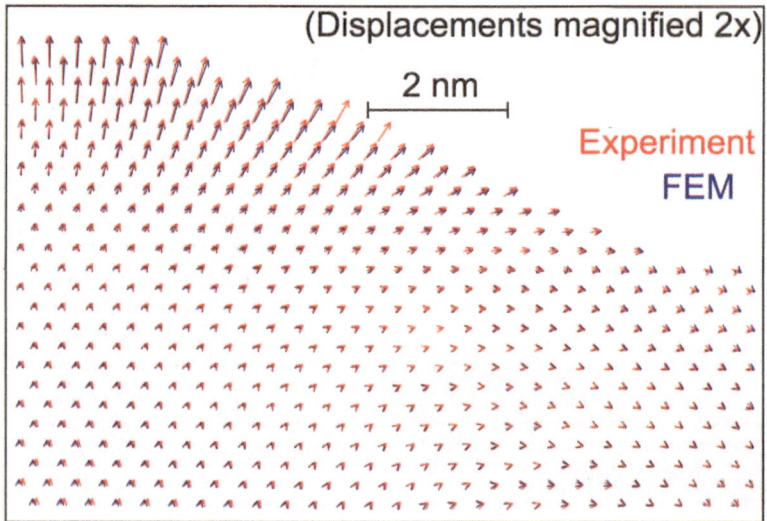

Plate 4. Part of the displacement vector field evaluated from Fig. 4.1 (drawn in *red*) and the corresponding field obtained by FE calculation (*blue*) as outlined in Sect. 4.3.2

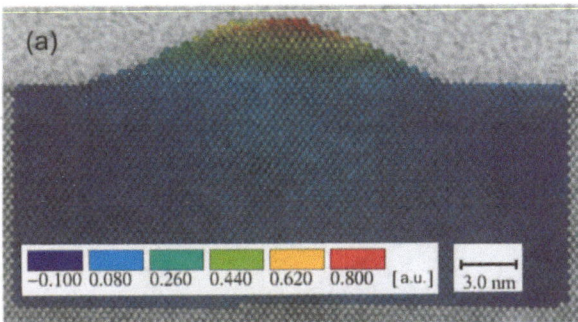

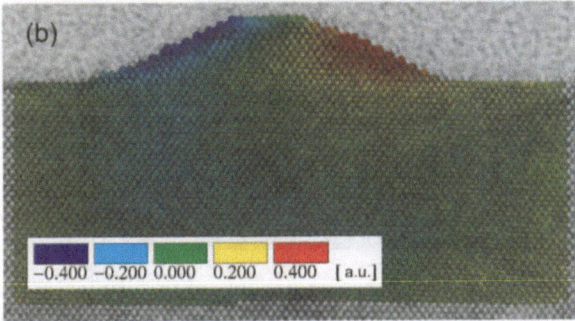

Plate 5. Color-coded maps of the components of the displacement vector field deduced from Fig. 4.1: **(a)** in the growth direction and **(b)** in the interface direction (a positive value indicates a displacement vector pointing to the *right*)

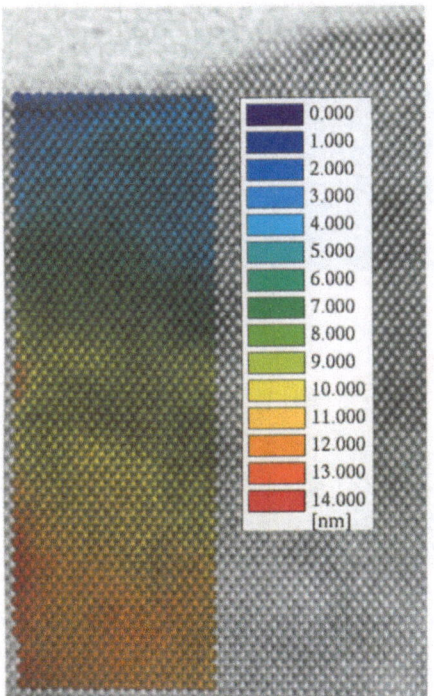

Plate 6. Color-coded map of the evaluated thicknesses

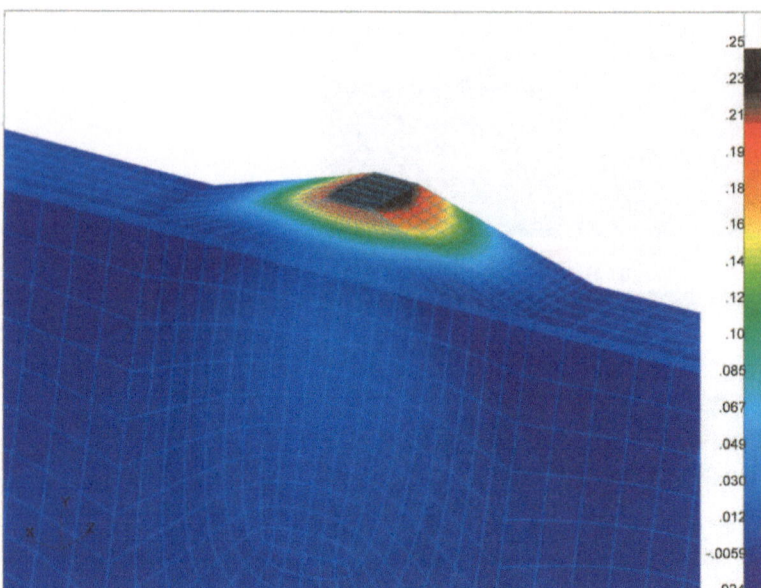

Plate 7. FE model with color-coded values of the components of the displacement vectors in the growth direction. The color-coded scale is given in nanometers. The *light-blue grid* indicates the finite elements

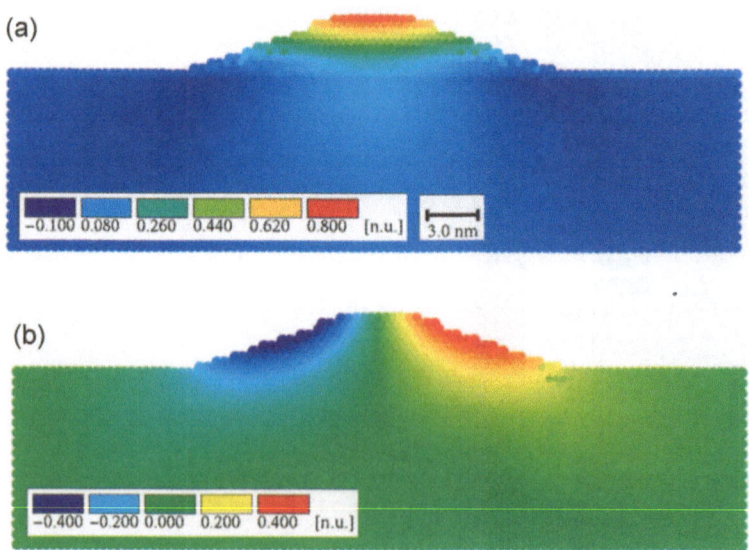

Plate 8. Components of the displacement vector field in (**a**) the growth direction and (**b**) the horizontal direction, evaluated from the FE calculation

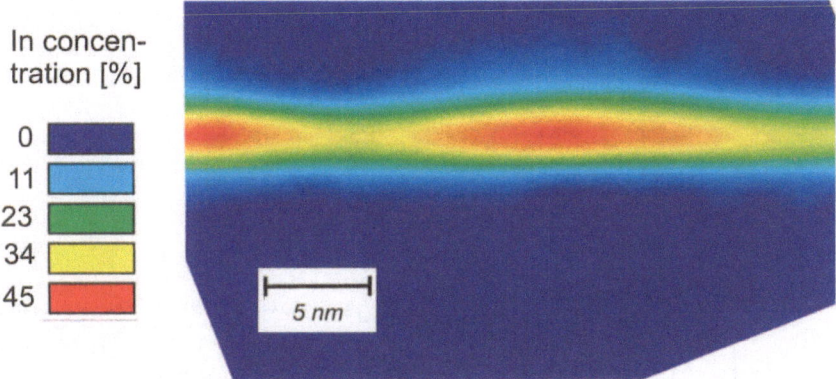

Plate 9. Color-coded map of the local In concentration

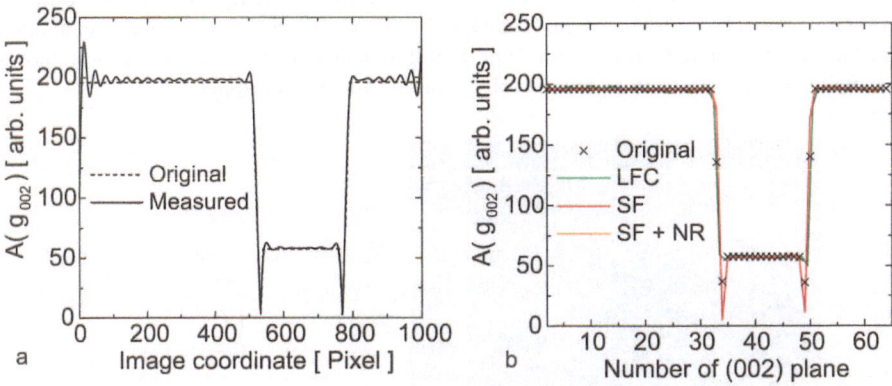

Plate 10. (a) Local amplitude of the (002) reflection $A(\mathbf{g}_{002}, \mathbf{r})$ plotted versus the pixel number in the vertical direction of Fig. 5.13a. The *solid line* shows the result of an evaluation by the method described in Sect. 5.3.1 and Fig. 5.9. The evaluation shows artificial oscillations that deviate from the true local amplitude of the (002) reflection, shown by the *dashed curve*. **(b)** Results of evaluations obtained by the following methods based on the decomposition of the image into image unit cells: computation of local Fourier coefficients (LFC, *green*) and model fitting in real space (SF, *red*). The *black crosses* show the true profile. The *orange curve* was obtained by model fitting in real space applied to the noise-reduced image shown in Fig. 5.13c. Note that the green and orange curves can hardly be seen because they are covered by the red curve

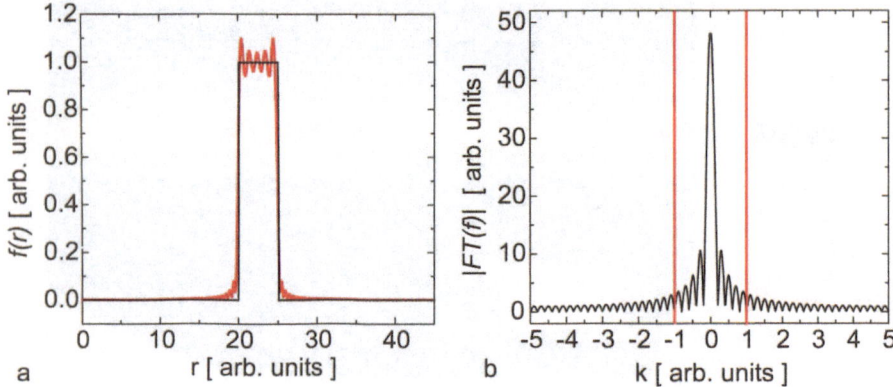

Plate 11. Graphs demonstrating the effect of filtering in Fourier space. (**a**) The *black curve* shows the original square function $f(r)$. (**b**) Amplitude of the Fourier transform of $f(r)$. Only the information inside the two red lines has been used for the inverse Fourier transform. The result is the *red curve* in (**a**), which shows artificial oscillations due to the filtering applied in Fourier space

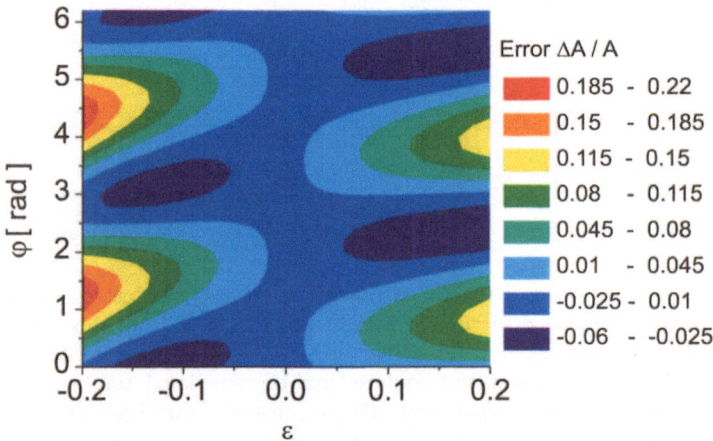

Plate 12. Relative error $\Delta A(\mathbf{g})/A(\mathbf{g})$ of the measurement of the local amplitude A of a reflection $\mathbf{g}$ by the local-Fourier-coefficient (LFC) method, induced by a relative difference ϵ between the unit cell size and the period of the lattice fringe image formed by a two-beam interference of the (000) and $\mathbf{g}$ beams

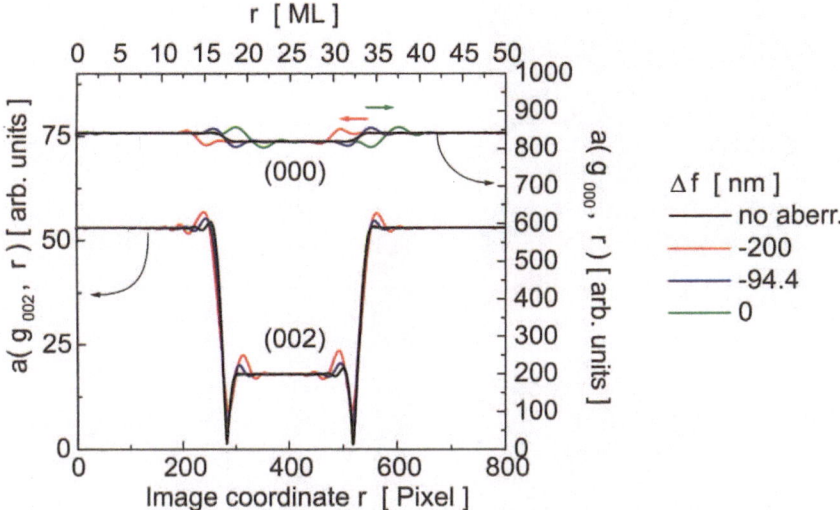

Plate 13. Profiles of the local beam amplitudes $a(\mathbf{g}_{002}, \mathbf{r})$ and $a(\mathbf{g}_{000}, \mathbf{r})$ plotted versus the spatial coordinate corresponding to the horizontal direction in Fig. 5.19. The amplitude profiles are given for defocus values $\Delta f = -200$, -94.4 and 0 nm, corresponding to Figs. 5.19b, c and d, respectively. The *three colored curves* were obtained from the aberrated wave function. The *black curves* correspond to the aberration-free wave function. The *arrows* show the delocalization of the (000) beam with respect to the nondelocalized (002) beam

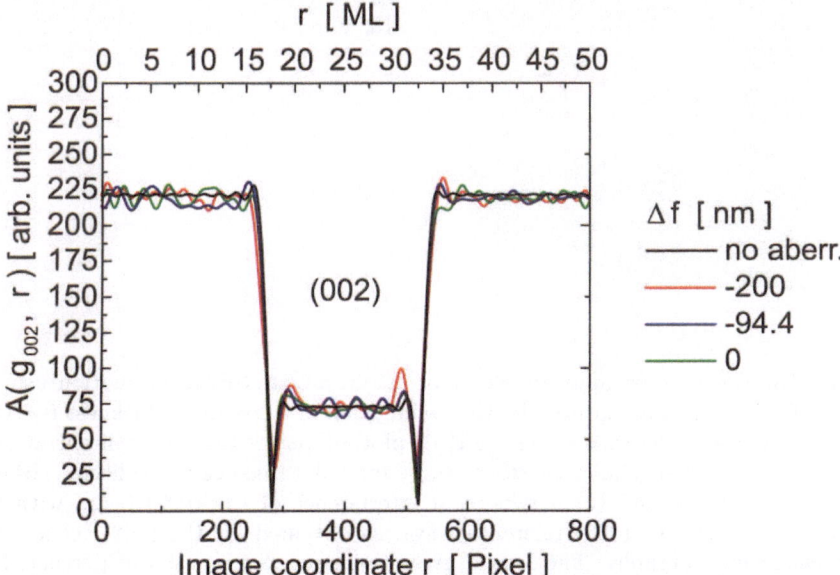

Plate 14. Profiles of the local amplitude $A(\mathbf{g}_{002}, \mathbf{r})$ of the (002) reflection for various defoci, analogously to Fig. 5.20

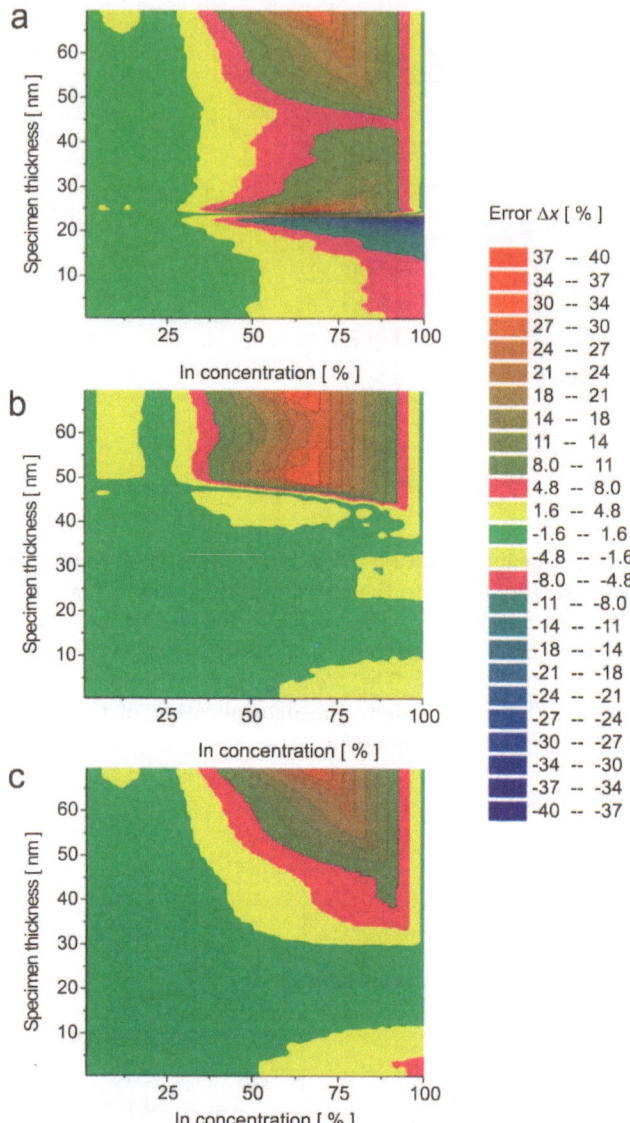

Plate 15. Color-coded maps of the error Δx in a composition evaluation using CELFA for $In_x Ga_{1-x}As$ induced by the assumption of a specimen thickness $t = 15$ nm and a phase angle $\varphi = 0$ (see (5.17)), plotted versus the true concentration x_{true} and the true specimen thickness t_{true}, for (**a**) three-beam condition, (**b**) a two-beam condition and (**c**) a two-beam interference of the (002) beam with a spatially homogeneous (000) reference wave, as obtained for the (002) reflection using electron holography. The legend gives the error Δx in units of percent. A detailed description of the imaging conditions is given in the text

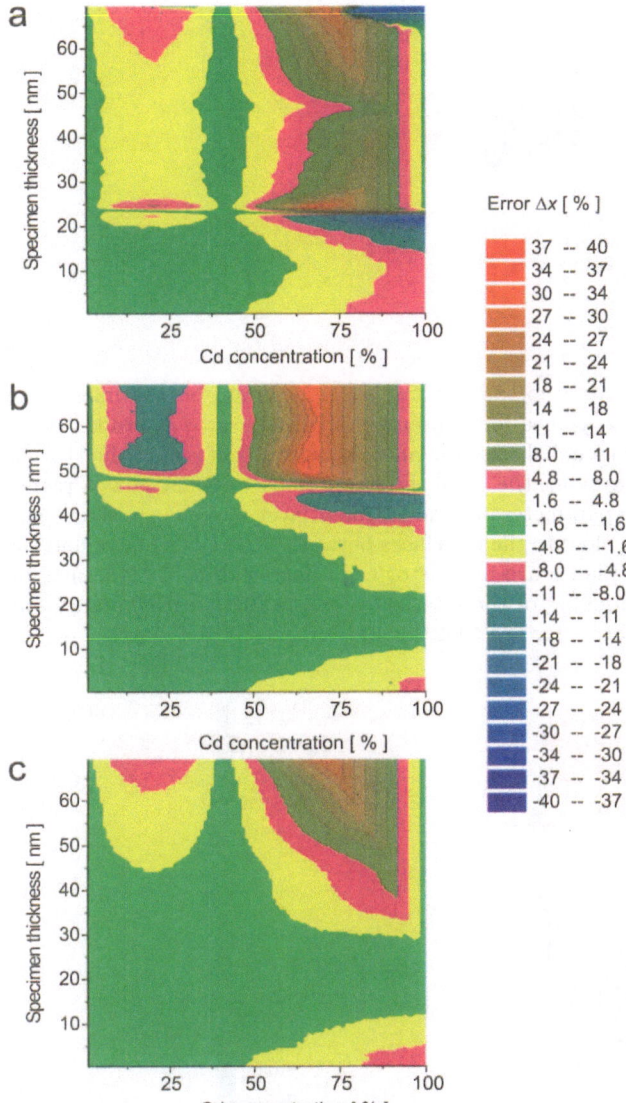

Plate 16. Color-coded maps of the error Δx in a composition evaluation using CELFA for $Cd_x Zn_{1-x}Se$ induced by the assumption of a specimen thickness $t = 15$ nm and a phase angle $\varphi = 0$ (see (5.17)), plotted versus the true concentration x_{true} and the true specimen thickness t_{true}, for (a) three-beam condition, (b) a two-beam condition and (c) a two-beam interference of the (002) beam with a spatially homogeneous (000) reference wave, as obtained for the (002) reflection using electron holography. The legend gives the error Δx in units of percent. A detailed description of the imaging conditions is given in the text

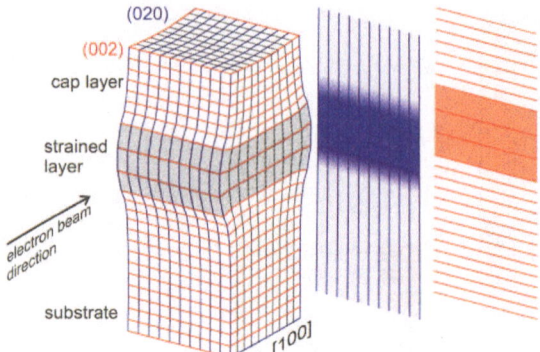

Plate 17. Schematic drawing explaining the effect of strain on the imaging of the (002) and (020) lattice planes for a pseudomorphically grown strained-layer heterostructure. The tetragonal distortion and its partial relaxation due to the TEM specimen being thin along the [100] direction leads to (002) lattice planes that have different lattice parameters in the strained layer and the substrate and show lattice plane bending close to the interfaces between the strained layer and the matrix. The (020) lattice planes are not distorted. Imaging using off-ZA imaging conditions leads to a broadening of the interfaces for the (020) lattice planes and the occurrence of strain effects for the (002) planes

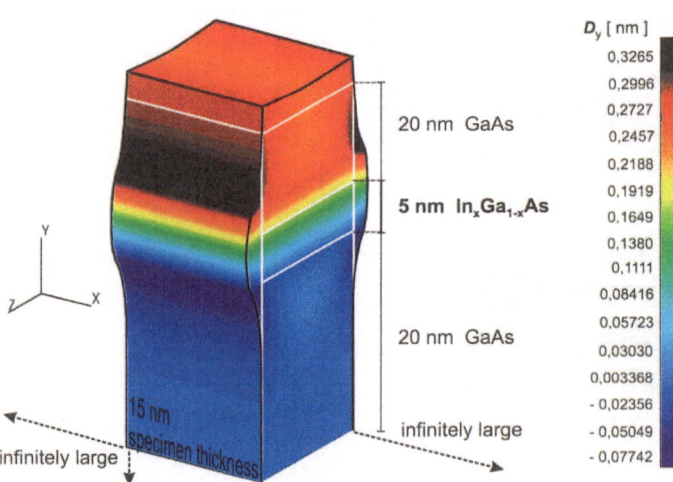

Plate 18. The *white lines* show the specimen model used for the FE calculation of the strain field of a 5 nm thick $In_xGa_{1-x}As$ layer grown on a GaAs substrate and covered with 10 nm GaAs. The specimen thickness is 15 nm. The *dashed arrows* indicate directions in which the model is assumed infinitely large; this was simulated by using appropriate boundary conditions. The distorted model shows the result of an FE calculation carried out for an In concentration of $x = 50\%$. For better visibility of the deformation field, the deformation is multiplied by factor of 10. The color code corresponds to the component of the displacement vector in the growth direction

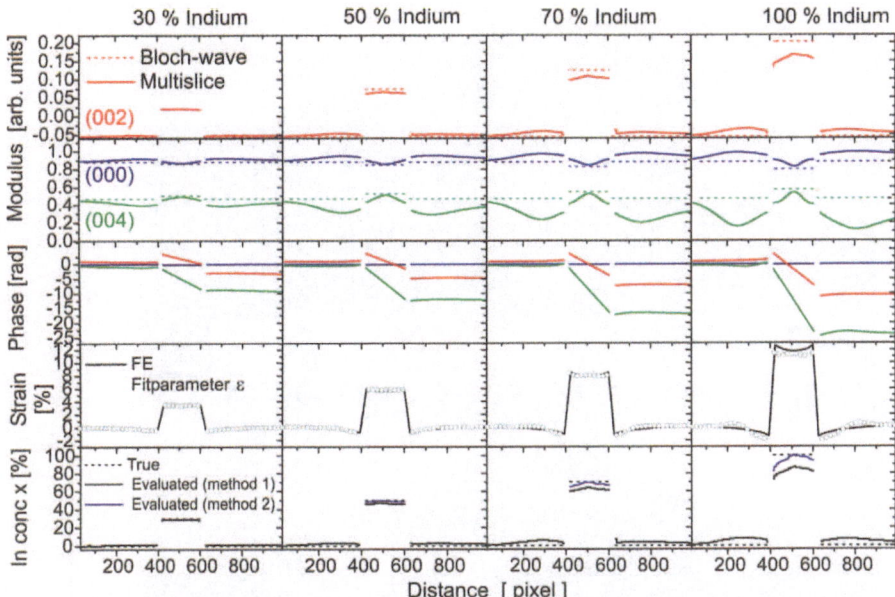

Plate 19. Graph showing the effect of strain on beam amplitudes and on the evaluated In concentration x for a 5 nm thick $In_xGa_{1-x}As$ quantum well buried in GaAs. Each column of graphs corresponds to one simulation, carried out for $x = 30, 50, 70$ and 100% indium. The *upper three rows* contain the moduli and phases (*solid lines*) of the (002), (000) and (004) beams, as evaluated from the simulated wave functions. The *dashed lines* show the beam moduli obtained from Bloch wave calculations, i.e. corresponding to an unstrained specimen. The *fourth row* gives the results obtained for the strain parameter ϵ of (5.41). The *solid curves* correspond to the strain obtained directly from the FE model by averaging along the electron beam direction. The *bottom row* shows the true concentration profile (*dashed curves*) and the concentrations evaluated from the amplitude of the (002) beam with (method 1) a set of Bloch wave data calculated using the lattice parameter of GaAs and (method 2) a set of data calculated using a lattice parameter corresponding to a pseudomorphically strained layer, where a full relaxation of the tetragonal distortion was assumed (thin-sample limit)

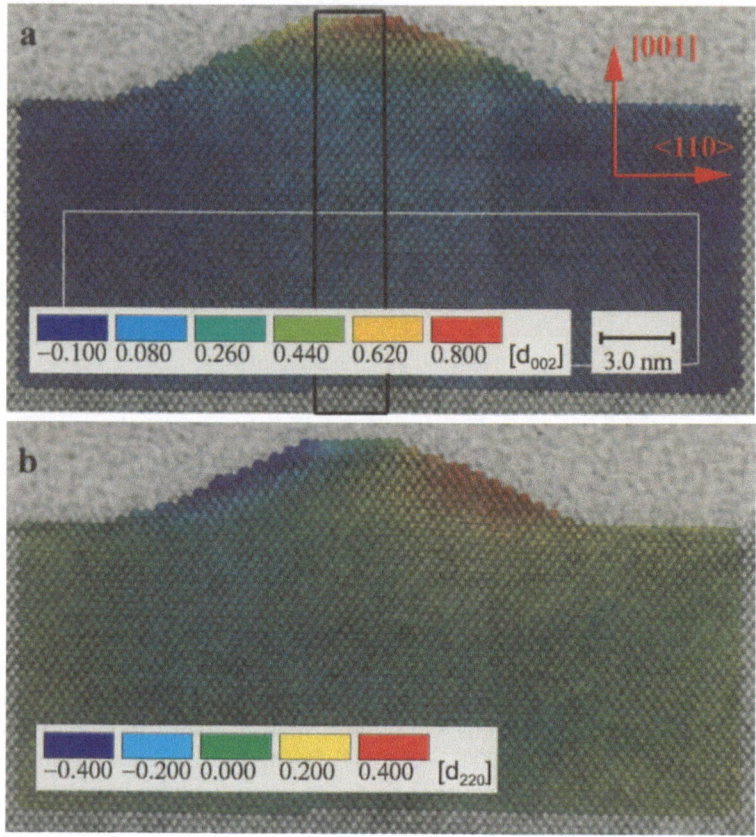

Plate 20. (a) Color-coded map of local displacements in the growth direction evaluated from a $\langle 110 \rangle$ HRTEM image of an uncapped sample with an $In_xGa_{1-x}As$ layer thickness of 1.5 nm. The displacement values shown in the legend have been normalized with respect to the averaged spacing d_{002} of the horizontal (002) lattice planes inside the reference region, marked by a *white frame*. The *black frame* marks the area that was used to adapt FE-simulated and experimental displacements. (b) Components of the displacement vectors parallel to the interface, in the same area. The displacement values have been normalized with respect to the average spacing d_{220} of the vertical (220) lattice planes inside the reference region

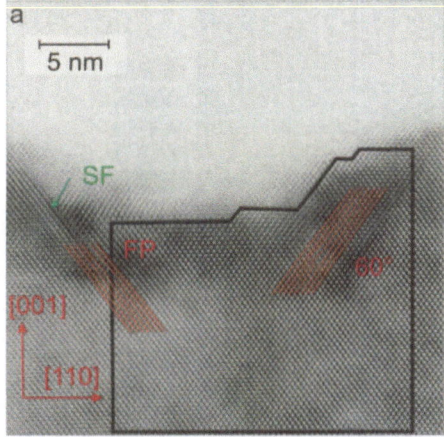

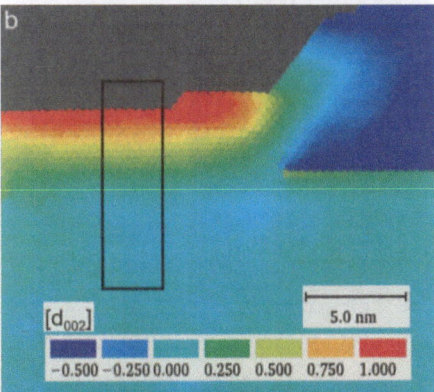

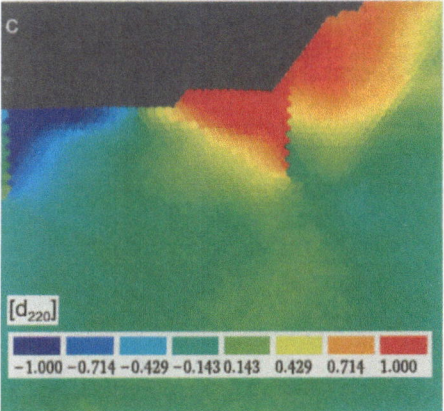

Plate 21. (a) HRTEM micrograph of a sample with a 2 nm $In_xGa_{1-x}As$ layer thickness taken in the $[\bar{1}10]$ ZA orientation, showing an island with a missing cap layer. The island contains a Frank partial (FP) dislocation and a 60° dislocation. The *red lines* help to identify terminating {111} lattice planes of the substrate. The *black frame* indicates the region that was evaluated by strain state analysis. The color-coded maps (**b**) and (**c**) show local displacement vector components in the growth direction and in the interface direction, respectively. The reference region was chosen to be inside the GaAs buffer. In (**c**), the abrupt transitions from *green* to *blue* and from *red* to *green* along two vertical lines at the *left* and *right* sides, respectively, of the island are due to (220) lattice fringes of the substrate that end at the dislocation cores. The *red region* corresponds to displacement vectors pointing to the *right*, and the *blue* region to displacement vectors pointing to the *left*

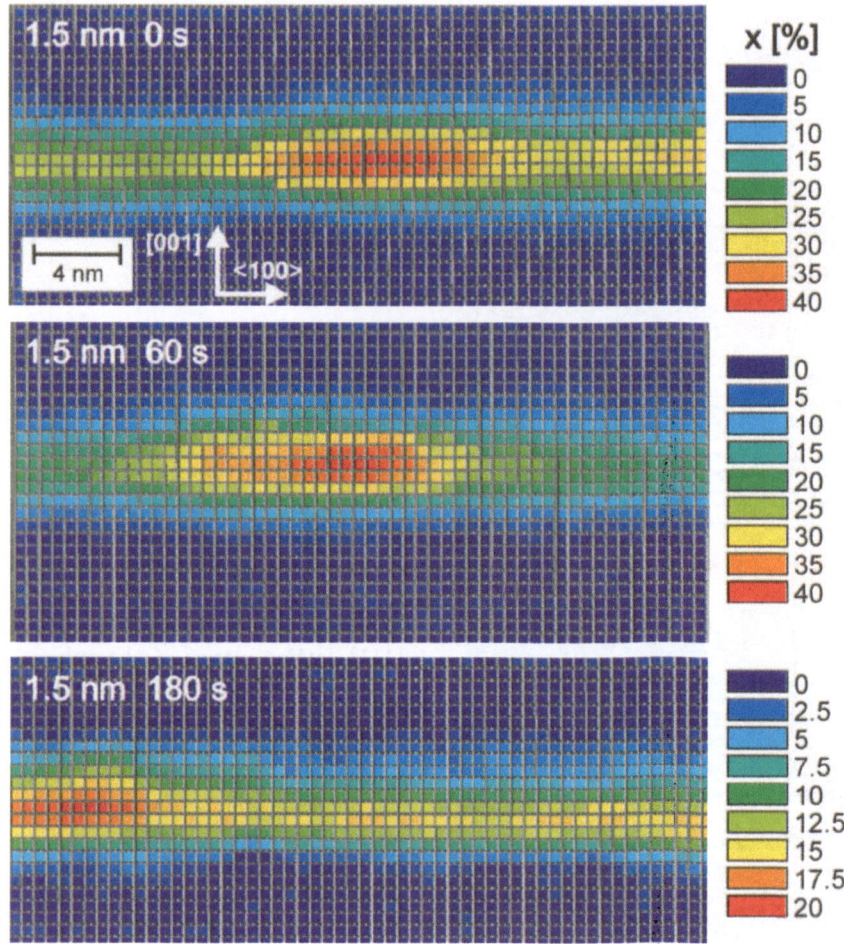

Plate 22. Color-coded maps of the local In concentration x evaluated by the CELFA method. Note that the *red* color corresponds to an In concentration of 20% for the 180 s specimen (*bottom image*), whereas it corresponds 40% for the *upper two images*

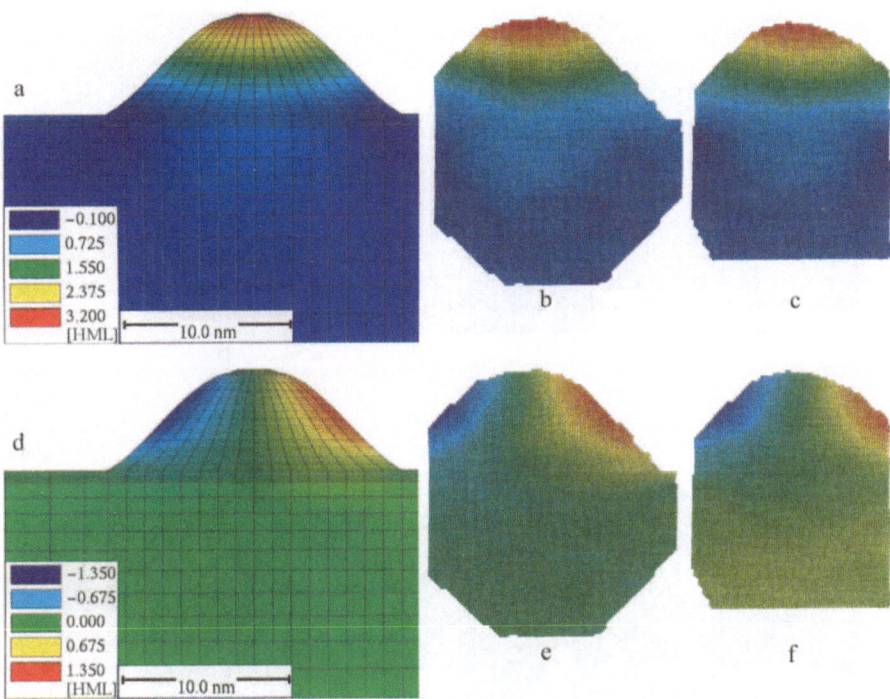

Plate 23. Color-coded maps of displacement vector components. (**a**),(**d**) Results of the FE calculation. The *upper row* of images shows the displacement vector components in the [001] growth direction and the *lower row* shows the components in the [010] direction (parallel to the surface). The legends give the displacement components in units of half an ML (HML). The color coding is kept constant for each of the two rows. (**b**), (**e**) Results for the island in sample C depicted in Fig. 8.3a. (**c**),(**f**) Maps of the island shown in Fig. 8.3b

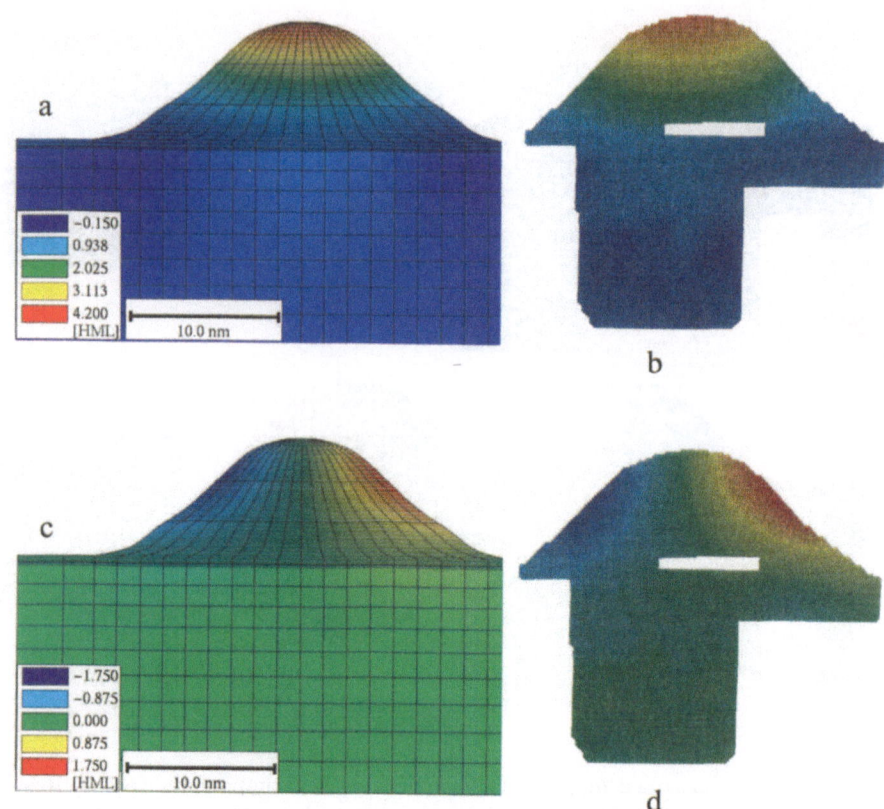

Plate 24. Color-coded maps of projected displacement vectors, analogous to those in Fig. 8.7. (b),(d) Results for the island depicted in Fig. 8.3 (sample D)

Index

Springer Tracts in Modern Physics

Springer Tracts in Modern Physics